D1309723

AN INTRODUCTION TO INDUSTRIAL RECREATION

Employee
Activities and Services

Prepared for
**THE NATIONAL
INDUSTRIAL RECREATION ASSOCIATION
CHICAGO, ILLINOIS**

Theodore B. Wilson
The University of Toledo

Robert Wanzel
Laurentian University
Virginia Gillespie
The University of Oklahoma
C.J. Roberts

wcb
WM. C. BROWN COMPANY PUBLISHERS
Dubuque, Iowa

Copyright © 1979 by Wm. C. Brown Company Publishers

Library of Congress Catalog Card Number: 79-50830

ISBN 0–8403–2031–0

Printed in the United States of America

C   402031   01

Contents

Foreword

The National Industrial Recreation Association has long recognized the need for an updated and well developed introductory text for Industrial Recreation. Although an excellent textbook was written by Dr. Jackson M. Anderson, there have been many changes and developments since its publication in 1955.

A small number of Association leaders laid the groundwork for a new textbook by beginning the collection of appropriate materials. Some industrial recreation professionals submitted manuscripts written especially for the purpose. After several months and a number of meetings, it was decided to secure the services of one or more professors in higher education to sift and evaluate the collected material, do additional research, and perform the actual writing of the textbook.

The present authors were selected and the collected material was placed in their hands. The Association also made available to them present and past issues of its official publication, *Recreation Management, Top Management Speaks,* "Keynotes," other Association publications and the results of Association research findings.

At the suggestion of the authors, the Association appointed a special textbook review committee of leading industrial recreation administrators and Association members. The committee reviewed and approved each chapter of the book before it was submitted to the publisher. The members of that committee, all Certified Industrial Recreation Administrators, were: Fritz J. Merrell, Olin Corporations; Roy McClure, Lockheed-Georgia Company; Miles Carter, Jr., McLean Trucking Company; Richard Brown, Texas Instruments, Inc.; William B. DeCarlo, Xerox Corporation; and Melvin C. Byers, NIRA consultant. The manuscript chapters were also reviewed by Michael Fryer and Patrick Stinson, successive Executive Directors of NIRA.

Upon the unanimous endoresement of the Review Committee, the National Industrial Recreation Association is pleased to accept the finished product. This new textbooks meets the need for an up-to-date source of information about the expanding field of industrial recreation. It will serve both the student and the practicing administrator as a valuable reference and guide. Because the book does set industrial recreation in its proper perspective and make its value manifest, we hope that it will be read by business and industrial managers and by government and civic leaders.

We are pleased that this new textbook brought to the attention of appropriate professors and administrators, will encourage institutions of higher education to offer more courses in Industrial Recreation—to the end that there be a continuing source of educated and specifically prepared professional administrators and assistants in this expanding and socially important field.

We commend it to you all. We are confident that you will find it interesting, informative, and useful.

For your Board of Directors
and the Text Review Committee

Fritz J. Merrell

Olin Corporation
NIRA President 1977–78

Preface

Surprise is still a word that is commonly associated with industrial recreation programs. Even though these programs had their beginnings in the 19th century, people today are still surprised to learn of an industrial recreation program at such and such an industry or business. Well, industrial recreation makes extremely good common sense and its time has come. The authors hope that this textbook will serve as the foundation for the further emergence and understanding of this field.

We are convinced that university and college departments of various disciplines would be well-advised to make a commitment to the inclusion of industrial recreation related courses in their curriculums. Through this intensive study, coupled with related research projects, the surprise element previously associated with industrial recreation will disappear.

Now, in ever increasing numbers, business and industrial leaders are demonstrating that they believe in the concept of industrial recreation. In this text we have outlined the potentiality for a multitude of benefits that can accrue to both management and labor through the inclusion of industrial recreation facilities, services, and activities. We hope that more firms, of all sizes, will examine these benefits and then implement industrial recreation programs.

The authors believe that this book will prove beneficial as a basic guide to professors and students in college and university courses and to professional administrators on all levels of organized industrial recreation service. We also feel that this text should serve as an excellent reference guide for a variety of recreation related organizations, not only in the Western Hemisphere, but in countries all over the world where industrial recreation has made an impact. We hasten to note that this book has been designed to serve only as an introduction and guide to the specialized field of industrial recreation. We anticipate that other textbooks will address themselves to various subfields.

Part I explores WHY industrial recreation exists. It examines the characteristics and background that industrial recreation developed within and what social forces were prevalent at the various times. It also expands on the theme that business must make a profit, but at the same time serve humanity. Management and labor need to relate better to each other and industrial recreation can be the catalyst.

Part II explains HOW industrial recreation can be implemented. By-laws, incorporation, and relationships of the organization to the company are discussed. Functional aspects such as finances, facilities, and legal matters are expounded upon; and a thorough analysis of program development, recreational activities, physical fitness, scheduling, and services is undertaken. Part II closes with a discussion of the professional staff involved and the professionalization of industrial recreation as a field.

Part III examines WHERE and WHEN industrial recreation began and is going. It looks at the origins, current status and future of industrial recreation and of its professional organization.

Finally, although individual recognition of all the large number of persons who have provided assistance is not feasible, we must express our special appreciation for the contributions of the following persons, all of whom are Certified Industrial Recreation Administrators (CIRAs):

To the best of our knowledge, Martha L. Daniell and Melvin C. Byers were earliest concerned about the need for a new and modern textbook and began the task of collecting the large mass of material which eventually came into our hands. Additionally, Melvin Byers served as the chief technical consultant during the writing of the book.

The five NIRA presidents who served during the preparation of this book all gave whole-hearted personal support to the project: Miles Carter, Jr., McClean Trucking Company; William B. DeCarlo, Xerox Corporation; Roy McClure, Lockheed-Georgia Company; Fritz J. Merrell, Olin Corporation; and Richard M. Brown, Texas Instruments Corporation.

Edward C. Hilbert, Battelle Memorial Institute, coordinated the project for the Association. He also gave his personal assistance, encouragement, and tactful and timely prodding, which was invaluable.

Much other valuable assistance came from a considerable number of industrial recreation professionals through manuscripts, letters, and telephone calls. To the extent identifiable, their material has been acknowleged in footnotes. We regret the likelihood that there are others whose contributions were unidentified.

We offer our sincere gratitude to the National Industrial Recreation Association as a whole. It has provided an abundance of financial, technical, and professional assistance. To an extent that few will recognize, the book is a cooperative product of the Association and the authors, with the Association deserving most of the credit.

We sincerely dedicate this book to NIRA and to its inevitable growth in numbers, strength, and social significance.

Acknowledgments

The National Industrial Recreation Association wishes to make special acknowledgment to a number of persons for their contributions to this textbook.

In May 1974, a group of eight persons met in Toledo, Ohio, and laid the plans which were to culminate in this book. Seven of those persons were notable leaders of NIRA: Miles M. Carter, then president; William DeCarlo, the president-elect; Martha M. Daniell, a past president; Melvin C. Byers, then vice-president and special consultant to NIRA; and A. Murray Dick, Director of Recreation of Dominion Foundries and Steel, Ltd., Hamilton, Ontario, Canada. Those leaders not only initiated plans for this book but assisted every step of the way in its development.

The eighth person at the Toledo meeting was C.J. (Posty) Roberts, Ed.D., Director, Office of Community Services, The University of Toledo. For the next two years Dr. Roberts was a factor in keeping the project alive. He spoke at NIRA meetings, collected additional materials, and served as the chief liaison agent with NIRA, the Wm. C. Brown Company Publishers, and the persons who did the actual writing of the book. His contributions are sincerely appreciated.

Other persons deserving of special recognition, and whose names are not mentioned in the text or footnotes, include: the 1976 Oklahoma Governor David Boron and Lt. Governor George Nigh; Mike Fryer, former executive director of NIRA; and Alan Vincent of Wm. C. Brown Company Publishers.

NIRA also feels special appreciation is due the outstanding executives whose words of wisdom appear in *Top Management Speaks* and are frequently quoted in the text. Recognition is deserved also by all the NIRA Board of Directors, who supported the project from the time of its inception.

Patrick B. Stinson

Patrick B. Stinson
Executive Director

PART I

1

Characteristics and Background

TOP MANAGEMENT SPEAKS*

In the past two decades, our society has been shaken by the winds of change. We have witnessed the decline of the work ethic, the springing up of the hippie movement and drug culture, the growth of a rootless society, and general dissatisfaction and disenchantment among large segments of our population.

Modern technology has made possible shorter working hours, more holidays, and consequently greater leisure time. One would anticipate that people would be happier and better satisfied than at any time in our history.

However, the opposite is true. The work ethic is no longer enough. With increased leisure hours we need programs that give people constructive, rewarding, and satisfying use of their discretionary time. We need a work-leisure ethic that brings people a sense of fulfillment and achievement in their daily lives.

I firmly believe that industrial recreation provides many answers to the problems of our society. We find this to be true in our own company. Most of our younger employees who are moving into positions of responsibility are those who came up through our recreation program. They are also proving to be active and interested citizens in the communities in which they live.

I am convinced that well-conceived industrial recreation programs are a tremen-

Earl T. Groves

dous stabilizing influence to our American way of life.

Earl T. Groves
Former President and Treasurer
Groves Thread Company, Incorporated

1

TOP MANAGEMENT SPEAKS

Recreational activities have been an invaluable asset of Battelle Memorial Institute from its earliest days. The work of a research and development organization such as Battelle calls for an intense and demanding creative effort, and our staff and management, from the outset, have recognized the need for recreation activities for physical fitness, relaxation, and personal expression.

Staff members have taken the initiative in establishing recreational activities, but the Institute has consistently lent its support, recognizing that the facilities and administrative services it provides are a valuable investment in the health and happiness of staff members and their families.

In an organization peopled heavily with scientists and engineers selected for their creativity, individual initiative has lead to a wide range of recreation activities. To name just a few, these include intramural sports, chess, model railroading, astronomy, and various musical groups. Recreational activities vary from one Battelle research center to another, as do the social events, but whereever Battelle people are—in the United States or Europe—there is ample opportunity for recreation for staff members and their families. It is a tradition that developed in our original laboratories in Columbus, and one that we have carefully nurtured at each new facility we have established over the years.

Sherwood L. Fawcett

We believe that a well-developed program of recreational activities enhances the creative environment we seek, leads to better working relationships, and stimulates the exchange of ideas—a particularly important element in interdisciplinary research. But much more important, we are convinced that recreation makes life happier and richer for all of us.

Sherwood L. Fawcett
President
Battelle Memorial Institute

Industrial recreation is much broader than and different from recreation in the usual sense of the word. In the beginning, however, let us focus on characteristics common to all forms of recreation.

Recreation and leisure, usually a necessary condition for recreation, are often misunderstood and undervalued in America. Thanks to the protestant ethic and our Puritan background, many Americans, especially the older ones, are ill prepared to appreciate leisure. The protestant ethic glorified work and our Puritan forefathers believed that "an idle mind is the devil's workshop" and that idle hands were the devil's servants. In a society where the opportunity for leisure abounds, many ignore it and others employ it unwisely. Robert Meynard Hutchins, educator and philosopher, expressed the opinion that civilization is less likely to be destroyed by nuclear bombs than by its inability to cope successfully with the leisure resulting from the peaceful application of nuclear energy. Socrates is quoted as saying, "Leisure is the greatest of all gifts," but George Bernard Shaw once remarked that "leisure is hell." Contemporary society finds the same dilemma.

For some persons leisure time is enriching and satisfying; for others it is boring and frustrating. We are far from the time when Americans had to "wrest a living" from the soil or from the sea, but social attitudes and values change less rapidly than the physical aspects of the environment. Yet attitudes and values do change—painfully.

Earl Groves, in the selection from *Top Management Speaks** which prefaces this chapter, includes the same impressions.

> In the past two decades, our society has been shaken by the winds of change. We have witnessed the decline of the work ethic, the springing up of the hippie movement and drug culture, the growth of a rootless society and general dissatisfaction and disenchantment among large segments of our population. . . .
> We need a work-leisure ethic that brings people a sense of fulfillment and achievement in their daily lives.

The changing of social values is always accompanied by some discomfiture. As the old, familiar values and social landmarks disappear, older persons, especially, tend to experience a feeling of being lost and insecure. Youth, on the other hand, are more inclined to accept change, or even to greet it with enthusiasm. The impact of change is the greater because change in one social value is always accompanied by change in other values or in the whole value system. As work has lost its halo of holiness, leisure has gained respectability.

Attitudes toward Play

In Puritan times, play by adults was equated with sin and debauchery. Even recently play has been considered frivolous and wasteful except for children. Yet today serious scholars consider play important for all ages. Sociologist Lloyd Saxton believes that "the contemporary adult who is unable to play is seriously handicapped in most of his activities."[1]

Most sociologists are convinced that the inability to play is damaging to one's relations with spouse, children, and even the general social and economic climate. Unquestionably, the inability

Top Management Speaks is a collection of statements reflecting support of industrial recreation by contemporary leaders. It is published by the National Industrial Recreation Association. Chicago: 1976. One or two selections from it preface each chapter.

of many adults to play is one aspect of the current generation gap. Play, however, is only one aspect of recreation, and it is the larger subject with which we are most concerned.

Definition Disputes

We might have begun with a definition of recreation except that there is no definition with which all authorities are in agreement. In defining "recreation," there are five general areas of disagreement. These concern whether recreation (1) is an activity, (2) is an end in itself, (3) must be voluntary, (4) occurs only during free time, and (5) must be socially acceptable and personally beneficial. Each of these areas deserves some discussion. All but the last of them are included in the following definition which seems so simple and uncontroversial.

> Recreation may be thought of as an activity voluntarily engaged in during leisure and primarily motivated by the satisfaction of pleasure derived therefrom.[2]

Whether an Activity

Whether recreation is necessarily activity embraces two points. The first is somewhat semantic—concerned with the meaning of words. Some authorities do not like the use of the word "activity" in the definition. To them, "activity" focuses upon what is being done rather than the effect upon the person performing the act. Those authorities consider recreation to be the resulting emotional condition. In a nutshell, the dispute centers upon whether recreation is the process or the product.

The second point is concern that "activity" implies motion. Recreation necessarily involves activity only in that it requires participation, as does pleasant meditation or the enjoyment of nature or music.

Whether Intrinsic

Whether recreation must have intrinsic value—be motivated by expected pleasure or satisfaction—or whether the same results may be achieved if the motivation is some external reward, poses another question. It may be asked whether the athlete is primarily motivated by the pleasure of running or jumping, for instance, or by the opportunity for competition—the chance to defeat someone else—or by the medal or prize. The answer probably varies.

Whether Voluntary

The question is whether recreation is necessarily voluntary. Does the fact that a particular activity is obligatory, more or less forced, necessarily mean that it is not recreation? Those who believe that recreation is not necessarily voluntary point out that there are times when we participate in something only because we must, but discover it to be an exhilarating experience. This is probably more often the exception than the rule, but it does happen. It is also true that some of the things we set out to do because we want to, turn out to be dull and boring. Circumstances do vary. "Playing" football as a professional often is work in the dreariest sense, though, for the small boys in the sandlot, playing football is marvelous recreation.

Whether Leisure Time

Is recreation only that which we do in our free time? There are those who think otherwise. What we read about Thomas A. Edison suggests that his work in the laboratory provided him with all the elements of recreation. Many people are sometimes so engrossed in their work and deriving so much pleasure and satisfaction from it that they are sorry when it is time that they must leave. Of course most persons, whether they like or dislike their work, weary of it and welcome the end of the day. It may be as safe to generalize about work as about leisure that some people enjoy it all of the time; some, most of the time; some, seldom; and others, never.

Whether Beneficial

Probably most professional recreation leaders tend to rule out by definition all activities which are not socially acceptable and personally beneficial. They take the position that such activities are not really recreation but mere stimulants or escape devices. There are those, however, who take the opposite position. The latter feel that the benefits commonly claimed for recreation can be achieved from barroom drinking, "grass" parties, the engaging of prostitutes, or any number of other activities which are illegal or generally held to be immoral.

In truth, there are some perplexing questions. For instance, could we conclude that playing poker or, for that matter, pinochle is recreation until midnight but not thereafter (the after-midnight players might be weary and less efficient the next day)? Or, is playing slot machines recreation in the States of Nevada or New Jersey, but not in states where such gambling is illegal? Shall we deny that some activity is recreation because *we* do not think it moral? Or socially acceptable? Or personally beneficial?

Definition Consequences

Definitions do matter! It is not necessary that we create a definition on which we can all agree; we cannot. It is that the elements of recreation definitions each have practical significance. The recreation director and staff need to be alert at all times to matters which affect the extent to which employees willingly participate in different aspects of the recreation program. As we shall note in chapter 6, there are a great many factors which influence participation.

Practical questions arise also from the academic dispute as to whether recreation may be engaged in only for its own sake or for extrinsic rewards. Decisions have to be made whether to stimulate participation by awarding prizes, or by other inducements such as company recognition. The opposite side of the coin is the question whether participation in any recreation project should be stimulated by negative peer pressure, that is, should anyone be led to participate in any activity or project because one's peers will disapprove of refusal?

It is amazing how many decisions have to be made about problems rooted in the issues posed by the definition of recreation. Granted, the director may not have to make decisions, but the director cannot avoid involvement. That is why thought should be given to the issues beforehand. It is better to be halfway prepared than caught completely unready.

The Director's Dilemma

Practical applications are the more difficult because there are no clearcut answers anytime, and the thorniest issues are those that are borderline. All that definitions can do is to provoke us to do some advance thinking and then help us to order our thoughts when relevant issues arise. Circumstances do alter cases, and issues will have to be resolved in the light of the special set of circumstances. Yet there need to be some criteria.

Whether the question involves wholesomeness, fairness, or finances, some criteria will be used. Should the criteria be the director's personal views or values? What the director judges to be wholesome or socially acceptable? In line with the company philosophy? Or believes to be best for the members? Of course the director can leave a problem up to the professional staff, if there is one. Or, each problem can be referred to a membership committee. But two, six, or a dozen persons can be as wrong as one; and each and all need guidelines.

The questions will come! One member will ask, "Is this activity appropriate for this organization?" Another will ask, "Is this really what we started out to do?" Still another will declare, "This is not what we should be spending our money for." There needs to be a set of guidelines that will be of assistance.

Actually the guidelines will not consist of definitions but rather plans that are described as purposes and objectives.

Purposes and Objectives

Every industrial recreation organization should give careful thought to what it wants to accomplish and why it exists at all. These things need to be put into writing for future reference. They are the guidelines which help to answer the questions which will arise. An example of such a statement is included here. It will help to explain the nature of industrial recreation and indicate how much industrial recreation differs from recreation in general.

The example is fictitious, and it is not intended to be a model. The authors neither defend the inclusion of any items nor the exclusion of others. Each recreation association's statement of purposes and objectives should emerge from its own individual circumstances and represent the values and goals agreed upon by its members. Chapter 4 will go into more detail about the development of purposes and objectives.

(Illustrative) Statement of Purposes and Objectives

This association (organization) is formed for the purpose of establishing a mutual benefit relationship which will further the interests of the members and the company. Its objectives are:

1. to provide recreational opportunities for the members and their families
2. to provide recreational facilities for the benefit of those concerned
3. to promote friendship and a climate of congeniality
4. to improve the individual members by the development of talents, skills, and knowledge
5. to provide release from working day frustrations and tensions
6. to improve employee morale and satisfaction with employment
7. to improve communication among the employees and between the employees and the company management
8. to encourage friendly cooperation
9. to prepare members for more responsible positions through the opportunity to demonstrate leadership ability, leadership training, and leadership experience
10. to improve the health and physical fitness of members
11. to provide beneficial services to the members, their families, and the community
12. to provide more cultural opportunities for members and their families
13. to provide greater educational opportunity for members and their families
14. to encourage members to develop skills and hobbies, and in other ways assist them to prepare for retirement

15. to serve company retirees by maintaining their contacts with associates at work, by providing retirees with recreational opportunities and recreational facilities
16. to further the interests of the company
17. to further the welfare of the community.

Again, the above statement of purposes and objectives is not intended for a model. It is not a copy of any known to exist, is not appropriate for all company associations, and may not be appropriate for any. On the other hand, it is not more ambitious or extensive than some industrial recreation associations do undertake and achieve. In that sense it serves to indicate the breadth and some of the other characteristics of industrial recreation. Thus it provides an easy transition to the topic of the nature of industrial recreation.

Industrial Recreation

Origins

Industrial recreation is recreation plus other activities and services for employees and related to their place of employment. It may be provided and financed entirely by the employer, entirely by the employees, or as a cooperative venture by both.

Almost everywhere that there are sizable numbers of employees, some of them will ultimately get together, informally at first, to organize an amateur sports team, have a picnic or other social gathering, celebrate Christmas, or arrange for flowers and condolences in the event of illness or death. Whatever may originally bring them together, the number and variety of activities in which they engage tends to increase with the passage of time. Ultimately such groups organize; but whether or not organized and whether or not recognized, industrial recreation exists almost everywhere.

A second way in which industrial recreation often comes about is through company largesse. In the beginning it may be a picnic which becomes an annual affair, the sponsoring of a sports team, or some other social activity. As those activities expand in variety and number of employees involved, there usually develops a formalized industrial recreation program.

A third common origin is the deliberate development of a new branch of a company where there is already a recreation program.

Sites of Programs

Initially industrial recreation developed in manufacturing industries, but subsequently it has spread to almost every type of employment. The only necessary factors are employers and employees. Today industrial recreation can be found also in mining, communication, finance, insurance, construction, distribution—retailing, wholesaling, and transportation—and in services—lodging, food, entertainment, health, education, maintenance, repair, and government. Evidence of this is to be found in the membership rolls of the National Industrial Recreation Association (NIRA).*

---

*Most of the time throughout this book we will refer to the National Industrial Recreation Association as NIRA (pronounced nīrà). That is the term most frequently used by members in referring to their national organization.

Program Emphases

Originally the emphasis of industrial recreation was on some kind of activity or social event. That is no longer true in most instances. Equal or greater emphasis is placed upon various kinds of services. These include, separately or in combination, (1) services for the employees, (2) services for the company, and (3) services for the community.

Services to the employees and, usually, members of their families include a wide variety of activities. These include opportunities for growth through travel tours, toastmaster clubs, and education, as well as the development of interest and skills in hobbies. Services also include discounts for tickets to commercial recreation parks and special programs, and discounts for commercial products such as automobiles, boats, tools, or toys. And for the members' families, the services may include facilities for family camping or picnicking, or blood banks, or day-care centers or summer camps for children. A complete list would require pages.

Services to the company include important contributions to communication, leadership, public service, and public relations—to name only a few of the benefits which will be elaborated upon in chapter 3.

Services to the community include assistance with special community projects such as fund raising or beautification; financial support and volunteer labor for disadvantaged children, the ill, and the elderly. This topic is enlarged upon in chapter 3 also.

Business Affiliations

Uniquely among recreation organizations, industrial recreation seeks the affiliation and co-operation of private businesses. This affiliation is open, without conflict of interest, and to the mutual benefit of the recreation group members, the recreation association, and the businesses concerned. Through such affiliations, the individual members benefit by the receipt of discounts from the enterprises and by the investigations of quality and dependability by their own organization. The recreation association receives information, special services, and significant financial support.

A Rose Is a Rose

There are those who would like to change the designation of industrial recreation because, in two respects, it is a misnomer. First, industrial recreation is much broader than recreation and for many members, perhaps the larger number, the services provided are more important than the recreational activities. Second, it is not limited to industry but, as noted, rapidly is being adopted by almost every other kind of employment.

Mel Byers, CIRA, former corporate director of industrial relations for Owens-Illinois Corporation, expresses the opinion that it would be better to think in terms of a mutual benefit society; and some employee groups use those words in the name of their organization, as, for instance, the Johnson (Wax) Mutual Benefit Association (JMBA). Richard Brown, Ph.D., CIRA, director of recreation for the Texins (Texas Instruments Employees), takes the position that industrial recreation "is something of a misnomer. 'Employee recreation services' more accurately describes our work."[3] Frank Flick, president of Flick-Reedy Corporation and one of the most articulate of proponents of industrial recreation, recommends a name change because, as he puts it, "What we are really talking about are all those leisure-time employee activities—whether recreational, educational, health-improving, or community-serving—in which both the company and the employees want to participate."[4]

We have mentioned these various ideas concerning the designation of industrial recreation only for the purpose of illustrating that those most involved, the real experts, recognize that in its application and function industrial recreation is much different from what the term itself would suggest. Yet there is a good deal of truth in the idea expressed in the poetic line, now almost an adage, which notes that a rose by any other name would smell as sweet. The term "industrial recreation" is so well-established now that changing it might cause more confusion than exists now. What is most important is that everyone understands the substance, that is, the real nature and characteristics of what is called "industrial recreation." Now that that matter has been explained, we are ready to move on to a discussion of the social forces which brought industrial recreation into being and affect it now.

Social Forces

Telesis is the theory that there is a purposeful and intelligent direction of natural and social forces to a desired end. Not everyone subscribes to the concept of telesis; but almost everyone will admit that there is something mysterious in the way in which society affects all its members. "Society," in the sociological sense, as it is used here, is nothing more than a group of people who have become more or less organized and share enough of values and behaviors that they consider themselves somewhat as a unit more or less distinct from other human units.

In a sense, society is only the aggregate, the sum total, of the human beings which compose it. Yet, in some inexplicable way, society acts as though it had a personality and being of its own. In some ways, society is independent of the individuals who compose it. Society affects the individual more than the individual affects society.

Society and Change

Society boths resists change and foments change. It is as though society realized change is inevitable but that too rapid change is destructive. Society responds constantly to internal and external pressures by changing its own shape and nature. It bends and adjusts in order not to be broken or destroyed. It does this by affecting the values and, hence, the behavior of the individuals which compose society. To paraphrase David Riesman and his fellow authors of *The Lonely Crowd,* society moves in some mysterious way to cause its members to *want* to behave in the ways in which they must behave if society is to flourish and continue. The point of this discussion is that industrial recreation in its present form is not an accident of history but the product of social forces. Some of those social forces we will want to identify. First we would like to acknowledge the complexity of social forces. Our conclusions are tentative, not proven.

Complexity of Social Forces

Social forces are somewhat like society. That is, they are mysterious and so complex that we can never be sure that we understand them. We can never be sure which is cause and which is effect, or, even, whether there is a cause-effect relationship. Conceivably neither of two phenomenon is cause or effect but both are the product of some third condition which in unrecognized.

For instance, it is a common observation that one of the "causes" of the increasing number of women in the work force today is the increase in the number of divorces. That observation may be true to some extent; but so is the reverse. That is, there are more divorces because there are more women in the work force. In the home, women have limited numbers of intimate contacts with the opposite sex; in the work force women and men have a multitude of intimate contacts.

Without moralizing, we can observe that the greater incidence of women and men working side by side and shoulder to shoulder has had obvious consequences. There has been a multiplication of unplanned heart-quickening glances. On occasion such glances have been followed by meaningful smiles. Smiles may have been followed by "accidental" touching of fingers at the water fountain. Thus, step by step, some working couples, one or both married, have been led to bedroom romances. The consequences of some such romances have been divorces and broken homes.

On the other hand both the increase of women in the work force and the increase in the number of divorces are products of other social forces, forces which are both related to and separate from the other two phenomena.

We have kept the complexity and mystery of social forces in mind and we ask that you keep them in mind also. To qualify every statement made about social forces would be too cumberson for us and too wearisome for you. In this spirit we discuss some of the social forces which affect the past, present, and future of industrial recreation. Some of those forces will surely be familiar to you, but some may not.

The Protestant Ethic

The Middle Ages historical period was characterized by relative stagnation of population, both in terms of numbers and in terms of geographical mobility. It was also accompanied by many other characteristics. Among them were an emphasis upon individual artisanship, and acceptance of extreme privileges for a small percentage of the population, and a dependence upon tradition.

The Middle Ages drew to a close as the yeast of change quickened. A rather rapid population increase led to the increased geographical exploration and the colonization of new lands, including America. All kinds of ferment—social, political, and religious—were increased by the contacts with the new world.

Society had the need for expanded productivity to accompany its population and geographical expansion, but it lacked the necessary capital. The gold booty of the Spanish conquistadores added some capital to the old world, but that was only a drop in the bucket compared to the need. The most significant source for capital was personal savings. It probably was not coincidence that the developing social value system made a virtue of hard work, frugality, thrift, and saving.

It probably was not pure coincidence either that in the countries where there there was the greatest social ferment there was also religious ferment. It was in the countries which experienced the Protestant Revolution that there was the greatest need for capital and, hence, the most emphasis upon hard work, thrift, frugality, and savings. Those "virtues" were associated with the Puritans; so those values are sometimes referred to as the "Puritan Ethic."

Another Misnomer

The social values of hard work, thrift, frugality, and saving were coincidental with and not the product of the Protestant Revolution nor of Puritanism. The religious leaders were preaching against the emphasis upon worldly goods at the same time as the merchants, entrepreneurs—a term used for the middlemen of developing capitalism—and ship owners were busily accumulating wealth and using that wealth to produce more wealth. The significant fact is that, whether or not misnamed, the accepted social ethic did provide the capital for the development named the "Industrial Revolution."

The Industrial Revolution

The industrial revolution began in England in the last half of the eighteenth century and, in America, in the first half of the nineteenth century. Briefly, it was characterized by the concentration of manufacturing in a few places, increased dependence upon technology, and mass production. It was necessarily accompanied by the concentration of labor and the growth of cities, both necessary conditions for the development of industrial recreation.

This discussion of the industrial revolution is admittedly simplistic, but it should not lead to wrong impressions of its force or importance. One impression to be avoided is the common assumption that the industrial revolution was a distinct point in time, like an explosion that happened and then ceased to be. The expansion of manufacturing and productivity has never come to an end. The expansion was more noticeable a century ago because it grew out of such a small base. It has become so large since then that further growth receives little notice. In actuality there probably has been increased acceleration of the rate of expansion because each technological invention breeds several more. One thing is certain: the tremendous output of material goods is greater now than ever before in the history of man. That output has led to changes which have prompted some persons to designate the modern period as the Age of Consumption.

The Consumption Society

Mass production mandates mass consumption. Without mass consumption the channels of commerce would become blocked and the wheels of mass production would have to be slowed or temporarily stopped. The age of consumption requires a different ethic and society has obliged. For instance, the attitude toward goods has changed remarkably even in two or three generations. From the outmoded value of treating goods carefully, handing them down from generation to generation and making things last, society has come full circle. The current attitude seems to be "use it up, wear it out, and throw it away." At opposite poles we see Grandfather, who received a gold watch when he became twenty-one and cherished it all his life, and contemporary hippies, who despise, or make a good show of despising, material goods.

Society has made a similar about-face in respect to human effort. Rather than making virtues of hard work and saving, society now has the greatest admiration for those who amass wealth with the least effort and in the shortest possible period of time. For the out-of-step individual who continues to work long and arduously, society has coined a term of scorn and derision: "workaholic!" In this social climate, the emphasis upon leisure is understandable. Another phenomenon which is understandable and which also has some relation to industrial recreation is the transition from rural to urban America.

Farm to City

The transition from the farm to the city is well-known, but it still deserves brief attention. In one respect it is similar to the expansion of mass production, that is, it is a continuing phenomenon. It is not something that happened once; it is still going on. It is also of greater magnitude than is often realized.

The facts are rather hard to marshal, for even the Census Bureau has changed its definitions during the course of time. Yet by any criteria, the trend away from the farms is unmistakable. In the first census of the United States ever taken (1790), only 5% of the population lived in what were then called urban areas. In the decennial census of 1970, 73% of the population lived in what is now defined as urban areas.[5] Of the remaining 27%, the larger part lived in small cities and towns. A study made in 1972 indicated that only 4.6% of the total of Americans were "farm population."[6]

The percentage of farm population will continue to contract, but more slowly, for some time to come. The individual, family-operated farm may finally give way almost entirely to the corporate owned and managed agricultural enterprise.

In rural settings people had few problems about recreation. Hunting and fishing were taken for granted. Most social activities revolved about churches and schools. Moreover, the demands of farming were so heavy that people were left with little leisure time. When farmers move to the city and become plant employees, it is quite a different story both for themselves and for their families. Today there is quite a margin of difference between the number of farm work hours and nonfarm work hours, but that was not always so.

Dwindling Work Hours

Until almost the middle of the nineteenth century, laborers commonly worked twelve hours a day six days a week (72 hours). By the beginning of the twentieth century, the average workday had been shortened to ten hours (60 hours per week). By 1940, the eight-hour day and the five-day week had become fairly commonplace. Today pressure is mounting for a shorter work week and that has already been achieved in some occupations.

Also, more usable leisure time has been provided by the stipulation that federal holidays shall be so calendared that there will be five three-day weekends each year. Vacations are growing longer. In 1972, over forty million people were working "under conditions of employment entitling them to three-week vacations."[7] Though lacking statistics, we know that four-week vacations are not unusual today.

Leisure at Home

As it happens, the greater leisure in the vocational environment is accompanied by a great deal more leisure in the home. Technology has worked wonders for the housekeeper. Prepared foods, ready-made clothes, disposable diapers, microwave ovens—these and countless other labor-saving devices have compacted a day's work into a matter of a few hours. We are indeed a long time and a long way from the period when it was said with some truth, "Men work from sun to sun, but woman's work in never done." Leisure has increased tremendously and is still growing. How it is and will be used is an important matter.

Uses of Leisure

The growing amount of leisure has many implications for industrial recreation, but they will not be discussed at this time. Rather, we are here concerned with the problems which accompany increasing leisure time. Though, remembering the earlier discussion of the complexity of social phenomenon, we will not suggest that the added leisure causes a growing number of social ills, such as alcoholism, mental illness, dependence upon drugs, and juvenile delinquency, we can hardly escape the suspicion that there is some connection. Beyond doubt, there is a correlation in point of time.

It is obvious that leisure can be used profitably or in ways which are self-destructive. Or, leisure can simply be wasted. Certainly there is something to be said for the opinion of Jay B. Nash that "to use leisure intelligently and profitably is the fianl test of a civilization."[8] Leisure can be used for wholesome play, for personal growth and fulfillment, and for service to others. These uses are the hallmark of the better industrial recreation programs, though we do not suggest that they are the only answer. Some answers will soon be needed because the effects of growing leisure probably will be intensified by the growing affluence.

Affluence

That the people of the United States, as a whole, are growing more affluent is observable and is supported by the available statistics. In 1967, the total purchasing power of the people of the United States was calculated by Conference Board statisticians to be 577 billion dollars. Of that total, it is believed that 40.8% was discretionary purchasing power—that part of the total purchasing power remaining after previous commitments and essential needs have been met, and over which the consumers have some choice. By 1974, the total purchasing power had grown to 1,037 billion dollars, of which the discretionary purchasing power represented 42.7%.[9]

The increase of affluence is even more starkly highlighted if one takes a longer period for comparison, as did Linden Fabin in "Affluence in the Family." Using 1960 as the base year, he concluded that by 1974, the average family in the United States was consuming half again as much goods and services. During that period of time, the number of affluent families increased dramatically. As summarized, the Conference Board estimate of "the amount of money available for 'the good life' . . . rose from about 30 billion dollars in the early 1960s to over 80 billion [1974] . . . and from less than 7% of disposable family income [discretionary purchasing power] to more than 12%."[10]

Despite many assertions and some rather wild "guestimates," no one really knows how much is actually being spent for leisure pursuits. Probably as reliable as any is the 1972 estimate of the Economic Unit of *U.S. News and World Report*. Its study, based upon data from the U.S. Department of Commerce and from a number of business associations, reached the following remarkable conclusions:

> The money Americans are now spending on spare-time activities exceeds the national-defense costs. It is more than the outlay for the construction of new homes. It surpasses the total of corporate profits. It is far larger than the aggregate income of U.S. farmers. It tops the over-all value of this country's exports.[11]

"Spare-time activities" includes athletics, travel, hobbies, self-improvement study, and all sorts of cultural events. The term includes the whole scope of even the best industrial recreation associations. Thus it hints at the present and prospective social and economic impact of industrial recreation. But for now we turn our attention to other social conditions which have impact upon industrial recreation programs.

Education

As recently as 1959, barely over half of the U.S. civilian population had four or more years of high school, but, by 1975, almost three-fourths had that much education.[12] The population with some college education increased from 15% in 1960 to 25% in 1975.[13] Such changes naturally affect the work force. A study by the Bureau of Labor Statistics discovered that in the year ending in March 1974, the number of workers with a high school education increased by three million while the number of nonhigh school graduates declined by one million. At that time, 68% of all workers had at least a high school education and 14% had four or more years of college.[14]

The increasing education and sophistication of the general population and the work force indicate good reason why there should be a growing amount of democracy, not only the political arena, but in all areas of life, including industrial recreation. The statistics also suggest the desirability of a larger number of educational and cultural clubs and activities than most industrial recreation programs are yet providing. Students can undoubtedly think of other implications of the increasing amount of education, though we move on to other significant characteristics of the population and work force.

Age Characteristics

Because of changing birth and death rates, the age composition of the population changes periodically. During the quarter of a century ending in 1974, those changes were of importance to industrial recreation. For brevity, the most pertinent facts are presented in tabular form. Table 1 shows percentages of the general population by selected age groups. Table 2 shows the percentage of the labor force by age groups. A comparison of the two is interesting.

TABLE 1
Age Groups as a Percentage of Total Population[15]

Age Group	1950	1974	Percent Change
18–24	10.6	12.7	+ 2.1
25–54	41.5	36.0	− 5.5
55–64	8.8	9.2	+ .4
65 and up	8.1	10.3	+ 2.2

TABLE 2
Age Groups as a Percentage of Labor Force

Age Group	1950	1974	Percent Change
16–24	19.5	24.4	+ 5.4
25–54	69.8	60.0	− 3.8
55–64	12.0	12.0	0
65 & up	4.8	3.1	− 1.7

Among the more important implications for industrial recreation is the indicated need to place greater emphasis upon hobby clubs which will stimulate interest and develop new skills as a preparation for retirement. This is also indicated in the statistics regarding longevity.

The Life Span

The life span of the U.S. population has been growing for a long time. In 1920, the expectation of life at birth was 54.1 years. By 1976, that expectation had increased to 72.8 years, a gain of almost nineteen years. Although most of that gain is the result of lowering the mortality rate at birth, there is a definite increase of life expectancy even after reaching adulthood. The life expectation of adults has increased by perhaps five years since 1940.

A not so pleasant fact is that the differential between the life expectancy of males and female grew during most of the period for which such statistics have been kept. In 1920, the differential between male and female life expectancy at birth was only one year, but by 1970 the differential had increased to 7.7 years. The trend toward the expansion of the differential may have stopped, however, for it has remained almost stationery for the last six years.[16] It is an interesting hypothesis that the differential is related to participation in the labor force and the stresses and strains that are the lot of both employer and employee.

Retirement

The aging of the population and the increasing life span both affect the number of retirees so that their numbers naturally would be increasing. Moreover, during periods of recession, when financial retrenchment becomes necessary, many companies offer special inducements to persuade employees to retire as early as ages fifty or fifty-five.

The problems of retirement are well-known. It does require financial adjustments which may be difficult. Emotional and psychological readjustment may be even harder, especially for males. The male retiree may feel that his usefulness is over, and his ego is seriously wounded. If, also, time hangs on his hands because he has nothing that he really wants to do, his life may be shortened and what there is left of it may be a burden to him and those around him. There is growing concern about retirees both because their problems are being better recognized and because they are increasing in numbers and will continue to increase for the foreseeable future. This is indicated by the mortality rates of recent years.

The Mortality Rate

The 1977 mortality rate of 8.8 per thousand population was the lowest ever recorded. Deaths from diseases of the heart declined by 2.1% and cerebro vascular diseases by 3.9% over the preceding 12 months.[17] Diseases of the heart are still the leading cause of death, but the deaths are occurring at older ages. The decline in the age-specific death rate for diseases of the heart and cerebro vascular system are accounted for in part by advances in medical and surgical care and, probably, in part by the growing consciousness of the American people of the danger of such diseases. In recent years there has been notably more concern about diet, regular exercise, and weight control. The statistics seem to constitute a strong argument for the continuation and expansion of physical fitness programs in industrial recreation.

Although the 1977 mortality rate was the lowest ever, it might have been still lower except for the relatively large increases in deaths caused by motor vehicle accidents, homicide, and suicide. Deaths caused by suicide in the 12 months ending in November 1977 were 7.7% greater than in the preceding 12 months. Suicide is of relevance to this chapter.

Suicide

Suicides in America are more likely to be males, middle-aged or older, and unmarried. A poor adjustment to retirement is apparently the cause or a contributing factor in many instances of suicide. This is indicated by the significant number of recently retired men who engage in self-destruction.

Suicide also seems related to conditions of employment. It is widely believed that an important cause of suicide is situations in which interpersonal relationships are distant and not group-oriented. Probably one reason for the increase of suicides is the growing tendency toward vast size of cities, factories, and other places of work, worship, or education. This is especially noted in manufacturing. The larger the number of other workers, the greater the chance that the individual will feel lonely and alienated. The larger the operation, the greater the likelihood that the individual employee will lack adequate personal contacts and so will be denied the psychic stability that can come from intimate interaction with meaningful others.

In the very large enterprise, the new employee tends to feel an isolated stranger, and it is difficult for some persons to get well acquainted with fellow workers on the job. Additionally many workers in large work forces know no person in the organization above their immediate supervisor. Orders reach them through a series of channels from upper management. This leaves the workers with little sense of security. Also it makes it less likely that the employee will develop the feeling of belonging that comes from identifying with the company.

These facts probably constitute one of the strongest arguments for the support of industrial recreation. Whatever else it may or may not do, the recreation organization almost invariably gives its members a strong sense of belonging. Above all, it is group oriented. It provides psychic security and emotional comfort. And we are not just talking about those persons who commit suicide. After all, suicide is the ultimate cry of despair! How many despondent, lonely, and miserable persons are there for every one case of suicide?

There is one other aspect of suicide about which we wish to comment. It is of interest in itself, but it also offers the convenience of introducing the next topic, which is the family. The majority of persons who commit suicide are males who are single. Whether they have never married, have been divorced, or are widowed seems not to be relevant. All that matters is that they are not married at the time. In America and throughout the world such statistics are kept, married men live longer than single men.

Families

The figures about suicide do not necessarily prove that married men are happier than single ones; but a number of relatively recent sociological studies all conclude that married persons are happier than the unmarried. The studies, of course, were concerned about the married and the unmarried groups as a whole, and the generalization about the groups is not intended to apply to individual cases.

Even so, the conclusions that married persons are happier than single ones are astonishing. There are many marriages (studies indicate at least 20%) which are in a state that has been facetiously labeled "holy deadlock." Spouses in such marriages do not sleep together, share few or no intimacies, and interact with each other as little as possible. For some reason, they simply tolerate each other. Add to the number in holy deadlock the number of marriages which end in divorce or desertion and you will have to conclude that about three out of every four marriages end unhappily. And then there is murder! Would you believe that *statistically* your chances of being murdered by your spouse are greater than your chances of being murdered by a stranger? If married persons, on the whole, are happier than the unmarried, what in the world must be the emotional condition of the singles? Small wonder that psychiatry and marriage counselling are booming!

Marriages

In recent years young persons appear to have become somewhat cynical about marital bliss. The marriage rate per thousand of the total population declined form 11.0 in 1972, to 9.9 in 1976, and then increased slightly—and probably temporarily—to 10.1 in 1977. The actual numbers of marriages, which peaked at 2,284,000 in 1973, were provisionally reported as 2,176,000 in 1977.[18] When one considers that the number of persons of marriageable age are at an all-time high, both the marriage rate and the number of marriages suggest some disillusionment with marriage. Yet at least one other factor may have an important bearing and make the previous statistics less meaningful. The last few years have seen a growing trend for young persons to live together without a marriage ceremony. For those under the age of 39, the number of such living arrangements is reported to have quadrupled since the 1970 census. In a large number of instances, perhaps the majority, the couple consider their living arrangement as a trial to test their compatibility for marriage. Thus it is likely that eventually many such couples will find their way to the marriage altar and swell the marriage statistics.

Divorces

Today most college textbooks on social problems no longer make any reference to the number of divorces. That is because of the current viewpoint that a "good" divorce is better than a bad marriage. The divorce rate is seen as only the manifestation of underlying social problems and thus not a problem in itself. Nevertheless, the divorce statistics are disquieting to most of us. The number of divorces began a trend upward in the late 1950s and have been in a steady, almost precipitous, increase since 1968. The rate of increase has slowed in recent years but annual increases still continue. The year 1977 marked the fifteenth consecutive annual increase in the number of divorces granted in the United States. Beginning with 1975, the total of divorces has exceeded one million each year.

The divorce rate increased from 4.0 per thousand population in 1972 to 4.4 in 1973, to 4.6 in 1974, to 4.8 in 1975, and reached 5.1 in the twelve months ending in 1977.[19]

Children of Divorces

A significant trend is that more and more children are in households broken by divorce. That may be because of the current view that children are better off in single homes of tranquility than in married homes where there is tension and quarreling. Not all states report their divorce statistics in a way that will permit the exact determination of the number of such children, but it is known that in 1974 well over 600,000 divorces involved one or more children.[20]

Women in the Labor Force

The number of women in the labor force has been increasing all this century but has accelerated since the 1950s. In 1975, there were 56,507,000 women in the labor force. The increase is largely the result of the larger numbers of married women who work outside the home. In 1940, only 16.7% of married women worked outside the home, but by 1975, as many as 45% were in the labor force.

Married women constitute an even larger portion of the females in the labor force. In 1940, married women accounted for only 36.4% of the female labor force. By 1975, the married women made up 62.2% of the female labor force.[21]

Working Mothers

Significant for industrial recreation is the growing number of women workers who have little children. A special study by the Census Bureau showed that the number of children with working mothers increased by 650,000 between 1970 and 1973. During the same period, the number of children in all families fell. In 1973, 29% of working women had at least one child under three years old, and more than half (67%) had one or more children between the ages of six and seventeen.[22] A later investigation revealed that in 1975 there were 6,371,000 working married women with at least one child under six years of age.[23]

Conclusion

All social forces, but particularly those discussed, indicate the need for programs which will contribute to the stability of the family. As we shall see in later chapters, industrial recreation increasingly does support the institution of the family. Before we move to those chapters, however, we concern ourselves with basic problems of capitalism. Those problems indicate a need for the partnership of industrial recreation and capitalistic enterprises.

Notes

1. Lloyd Saxton. *The Individual, Marriage, and the Family,* 2nd. ed. (Belmont, Cal.: Wadsworth Publishing Company, 1972), p. 358.

2. Charles K. Brightbill and Harold D. Meyer. *Recreation: Text and Readings.* (New York: McGraw Hill Book Company, 1963), p. 10.

3. "How to Start an IRC in Your Community." *Recreation Management* [hereafter, *R.M.*]. April 1976, p. 26.

4. *The Untapped Potential: Industrial Recreation.* (Chicago: National Industrial Recreation Association, n.d.), p. 5.

5. Bureau of the Census. *Statistical Abstract of the United States,* 96th annual edition. (Washington: Government Printing Office, 1975), p. 26.

6. Peter Wohl and Robert H. Binstock. *America's Political System.* (New York: Random House, 1975), p. 245.

7. "Leisure Boom: Biggest Ever and Still Growing." *R.M.,* June/July 1972. Reprinted from *U.S. News and World Report,* 17 April 1972.

8. Jay B. Nash. *Philosophy of Recreation and Leisure.* (Dubuque, Ia.: Wm. C. Brown Company Publishers, 1953), p. 20.

9. *The Conference Board Statistical Bulletin,* No. 12. (December 1975), p. 5.

10. *Conference Record.* October 1975, p. 49.

11. "Leisure Boom." *R.M.,* June/July, 1972, p. 23.

12. *Statistical Abstracts,* 96th ed., p. 142.

13. Linden Fabin, "Affluence in the Family." *Conference Record.* October 1975, p. 56.

14. *The Wall Street Journal,* 26 March 1974.

15. *Statistical Abstracts,* 1975. Table 1, p. 6., Table 2, p. 344.

16. U.S. Department of Health, Education, and Welfare. "Advance Report, Final Mortality Statistics, 1976." No. 12, 30 March 1978, p. 2.

17. H.E.W. "Provisional Statistics: Births, Marriages, Divorces, and Deaths for 1977." No. 12, 13 March 1978, pp. 2–3.

18. Ibid., p. 2.

19. Ibid.

20. H.E.W. "Monthly Vital Statistics, Advance Report: Final Divorce Statistics, 1974," p. 3.

21. *Statistical Abstracts,* 1975, p. 347.

22. The Wall Street Journal, 14 May 1974.

23. *Statistical Abstracts,* 1975, p. 347.

2

The Dilemma of Capitalism

TOP MANAGEMENT SPEAKS

Recreation has been part of the Johnson Wax employee's way of life since the founding of our company in 1886. For example, it was the custom of my great-grandfather to give the employees a picnic every summer in the backyard of his home. And, at the turn of the century, men's and women's teams were formed to compete in basketball, baseball, softball, and golf.

Today, our 3,000 U.S. employees may participate in more than seventy different sports, clubs, special events, and services. However, the objective of our recreation program remains the same: to promote employee loyalty, fellowship, high morale, and physical and mental development.

Recreation is a great equalizer, a good ice-breaker, and often an incentive for employment, combatting absenteeism and turnover. Intangible as they may be, both the individual and corporate benefits are many.

We believe the recreation program at Johnson Wax is successful for several reasons. Above all, we get our employees involved in the planning and administrative aspects of the various activities. Even though we have a professional recreation staff, we consider this degree of employee involvement to be vital. We want and encourage our employees to be in on the decision-making in order that they may feel the recreation program is truly *their* program.

In addition to this committee involvement, we regularly employ the use of written

Samuel C. Johnson

evaluations/questionnaires in which participants are asked to give, anonymously, their opinions and suggestions concerning a particular activity. Our recreation program is flexible; we are not afraid to make changes or to break traditions.

I am happy to say that we have also been a forerunner in the area of providing a recreation program for our retirees. At Johnson Wax, we do not believe our relationship with a retiring employee ends with a party and the traditional gold watch. We work hard to keep a continuing relationship with the people who contributed so much to the success and growth of our company. We want our retirees, as well

19

as our employees to feel they are part of the Johnson family.

I would like to leave with you an excerpt from a profit-sharing speech made by my grandfather back in 1927. Although the message is fifty years old, it speaks of my sentiments so well:

"When all is said and done this business is nothing but a symbol . . . in a very short time these lively machines will become obsolete and the buildings for all their solidity must some day be replaced. The goodwill of the people is the only enduring thing in any business. It is the sole substance and the rest is shadow!"

Samuel C. Johnson
Chairman and Chief Executive Officer
Johnson Wax

Paul P. Davis

TOP MANAGEMENT SPEAKS

In many unionized industries, employees have become polarized because of contract restrictions that tend to prohibit intermingling. McLean's recreation program acts to prevent this. In fact, interaction is encouraged. A winning foursome in a recent golf tournament included a long-haul driver, a city driver, a supervisor, and a dock foreman. Last winter a bowling team included a terminal manager, a rate clerk, two city drivers, and a dock worker.

This broad base of participation from all areas of the McLean work force, with its by-products of good employee relations, and better morale and work attendance, has spurred complete management support of the entire program.

No longer do we consider recreation a fringe benefit. It is the catalyst that produces healthy, vigorous, dedicated employees who are an asset to McLean and outstanding citizens in their communities.

Paul P. Davis
Chairman of the Board and Chief
Executive Officer
McLean Trucking Company

This chapter has two messages, one addressed to professional industrial recreation personnel; one, to management.

To professional industrial recreation personnel: *You need the company. It deserves your loyalty and support.*

To management: *You need a good industrial recreation program. It is a sound investment.*
This chapter has two themes:

> Business must make a profit! Otherwise it perishes.
> Business must serve humanity! Otherwise it perishes.

Two items from *Business Week* of 9 August 1976 set the tone for the chapter. One is: "Reversal of Policy: Latin America Opens the Door to Foreign Investment Again." This story is about the Latin American countries which had expropriated the property of foreign "greedy capitalists" who had "exploited" their country for profit. Subsequently, seeing their economies languish, their industries slowing, their unemployed workers growing disaffected, and the rural poor growing poorer—seeing all that, they had a change of heart and were seeking to induce foreigners with the capital and the "know-how" to return and get the economy moving again. Jose Martinez de Hoz, Argentina's minister of economy said it all: "The real driving force in the economy is private enterprise. Profit is something that must be recognized as essential to the functioning of a market economy." [p. 35]

The second item is: "Can TWA Afford Its High Cost of Flying?" Excerpts are representative:

> Transworld Airlines is crawling back from the edge of an abyss once again. . . . Without high profits it won't be able to replace its aging fleet. . . .
> In short TWA is faced with a massive reequipment program, probably in five years. It cannot finance such a program unless it is solidly profitable. [p. 60]

The Profit Motive

It must be recognized that a primary responsibility of every business and industry is to make a profit in order that it can survive and provide society with needed goods and services. Society's needs grow constantly and rapidly—not only as the result of expanding population, but also in order to honor the contemporary ethic of extending "the good life" to all segments of the population.

There is a social need for the intricate and complex organization of production—for the goods and services themselves, and for remunerative jobs for the population.

If society as we know it is to be continued, there must be the accumulation of capital

to build new plants to meet new needs
to replace outworn or outdated equipment
to realize the fruits of new inventions
to provide for research and development
to keep the United States abreast of other countries
to improve the quality of life
to attract more capital.

Criticized as it is, the profit motive remains the strongest economic force the world has yet discovered. Idealism is wonderful, but people work harder for themselves than for the general advancement of society. From the founding of Jamestown in 1607 until today, every communistic experiment in America has floundered on that principle.

The profit motive is a foundation block of democracy. The free market functions as a superb cybernetic mechanism. It continuously monitors consumer desires, makes production decisions, and readjusts itself without the necessity of a central regulating bureaucracy.

Capitalism on the World Scene

Curiously the attack on the profit motive in America grows even as the non profit motive disproves itself throughout the world. A few nations should suffice as examples.

Italy. Italy moves ever further to the left. In the 1976 elections the communist party made such inroads that it had to be accepted into partnership in the government. Yet even as communism grew, the national economy weakened and stood on the verge of disaster. It might have gone over the brink had not Western capitalist nations provided millions of dollars to support the Italian lira. In 1978, it remained one of the weakest of European economies, aghast at each new revelation of public and private corruption and unable to respond adequately to the inhumane thrusts of the Red Brigade.

England. In England, the economy worsened steadily as the nation moved to nationalize more and more of its industries. By 1976, the once powerful empire builder and Mistress of the Sea was plagued by disastrous inflation and a balance-of-trade deficit that threatened its financial solvency. Subsequently, its trade deficit narrowed by the inflow of North Sea oil, England in 1978 is under the ominous clouds of continued inflation, declining production capacity, unemployment, and labor turmoil. It may well be that the dangers facing England today are less dramatic but no less serious than the threats posed by the Spanish Armada and the German luftwaffe.

China. China is in the beginning of the inevitable, slow process of watering down its communist idealism in the interest of pragmatism. In 1976, it began to expand its commerce with the capitalistic nations with two objectives: to feed its millions and to import the capitalistic technology and equipment to modernize its productive resources. In 1978, China defiantly faces, almost literally at bayonet's length, its former communist friends: the U.S.S.R., the pristine communist Albania, and Viet Nam, which it helped to rescue from the dangers of capitalism. Internally, it is torn between its revolutionist-type communist idealists on the one hand and on the other its current pragmatic leaders who see the value of personal incentives for production as well as the need for the technology possessed by capitalist nations.

The Union of Societ Socialist Republics, the show place of communism. The Soviet Union now pays only lip service to pure communist theory: from each according to his ability; to each according to his need. In its economic policies, Russia has acknowledged the power of the profit motive. By 1978, there was a tremendous difference between the incomes and prerogatives of the commissars and industrial elite on the one hand and those of the factory workers and peasants on the other.

The agricultural workers themselves prove the power of the profit motive. In the Soviet Union about 4 percent of the land is left out of the collective farms. The individual farmers are permitted to use that land to grow crops for sale. That 4 % of the land produces, by value, about 25 % of Russia's total agricultural production.

In 1975, Nikita Khruchev replied to the charge that the communist leaders were moving toward capitalism: "Call it what you will, incentives are the only way to make people work harder."[1]

The Attack on Capitalism in America

On the American scene the attack against the capitalistic system mounts. In 1976, the attacks on business ranged widely. The ideological liberals said that the profit motive is destructive to society and that business must put social responsibility before profits. The Old Left (the traditional communists), with seemingly less conviction, argued for communism. The New Left, though lacking any solution of its own, insisted that the capitalistic business system was evil and must be destroyed. The New Left, little noted and underrated, is significant and deserves some attention here.

The New Left

Almost unheard of before 1962, the Students for a Democratic Society (SDS) had grown by 1968 into "the largest and most influential radical group in the country, with more than 7,000 dues-paying members and 400 campus chapters."[2]

The Port Huron Statement

The year 1962 marked a memorable event for this group composed largely of college students. In that year some of the members gathered at a retreat at Port Huron, Michigan, to discuss a statement of principles for the organization. As later revised, that statement of principles became the most significant document of the New Left. It is known as the Port Huron Statement and more than 100,000 copies eventually were distributed.

Some of the introductory sentences illustrate the idealism, the rationality, and the persuasive power of the originators.

> We are the people of this generation, bred in at least modest comfort, housed now in universities, looking uncomfortably to the world we inherit. . . .
> . . . We began to sense that what we had originally seen as the American Golden Age was actually the decline of an era. The worldwide outbreak of revolution against colonialism and imperialism, the entrenchment of totalitarian states, the menace of war, overpopulation, international disorder, supertechnology—these trends were testing the tenacity of our own commitment to democracy and freedom and our abilities to visualize their application to a world in upheaval.[3]

New Left Issues

At the time of the original Port Huron Statement, the chief issues were national: colonialism, the Cold War, and nuclear disarmament. The eventual strong thrust of the organization at the evils of the business system was only hinted at then.

By 1968, the more committed and ideological SDS leaders had fully developed the logic of their attack on business as a system. The rank and file of college students were even then more incensed by militaristic activities on university campuses. Most of the student demonstrations of the 1968–69 academic year seem to have been directed toward the removal of military research projects and the ROTC from college campuses.

The Ideological Split

Controversy grew within the SDS movement about principles, priorities, objectives, and methods. The SDS ousted the Progressive Labor Party faction—the students who wanted to work for world communism. The remainder of the SDS split into two groups: the Revolutionary Youth Movement I (RYM I), known as the Weatherman faction, and RYM II, a mored moderate group. The Weathermen faction turned to violence and terrorism, thus discrediting forever, for most Americans, the entire SDS movement.

Nevertheless, the RYM II dedicated ideologists continued writing and speaking their advocacy of the destruction or radical revision of the American business system as a system. Their quarrel, it should be noted, was not with business or governmental leaders as individuals but with the business system as a system. That they did not go unheard is attested by a 1968 article in the *Harvard Business Review*, "What Businessmen Need to Know about the Student Left." It warned that the New Left was "a relatively small but hard core of young student revolutionaries aiming at nothing less than the destruction of U.S. society in general and the Business Establishment in particular."[4]

Actually the New Left was not the first to identify business as an evil system. In 1912, Woodrow Wilson lamented: "We are all caught in a great economic system which is heartless."[5] The New Left, however, not only identified the business system as evil, but they mounted a strong, sustained, and rational attack on the system. Businessmen were not moral monsters as much as simply parts of the immoral system.

According to the New Left, business dominated government for its own pernicious ends, perpetrated social injustice, widened the gap between the haves and have-nots, enslaved the weak and defenseless—there were not the faults of businessmen as persons but simply the results of businessmen's doing of their job. That is why the whole system had to be destroyed or radically altered.

New Left Methodology

Although sometimes employing the exaggerated rhetoric common to radical reformers, the authors of the New Left reflected their college backgrounds. Many of their articles were scholarly and well-researched. They used statistics rather than generalized, unsupported statements. They often documented their charges with careful case studies.

They were studious in the areas of their concern, effectively using the material from muckraking journalism, socialist literature, and the publications of respected social scientists and historians. They quoted from prestigious, business-oriented journals such as *Fortune* and the *Wall Street Journal*. And they were intelligent and logical in the use of their material. Much of what the New Left wrote had the ring of truth because it was true!

The Lasting Significance

The impact of the New Left was somewhat lessened by two factors: (1) the excesses of the Weatherman faction tainted all student radicals by association in the public mind; and, (2) for all the brilliance of their attack on the business system, they were never able to develop a sound alternative.

Yet in spite of their problems and their small numbers they have been and are a significant force in American thought. There are at least four important reasons.

1. They initiated a rational attack on the business system. Their thrust was sufficiently enlightening and persuasive that it has subsequently been adopted and enlarged upon by recognized scholars with impressive credentials.

In January 1975, "seven Nobel prize winners, including Economists Gunnar Myrdal and Kenneth J. Arrow, signed a declaration condemning Western Capitalism for bringing on a crisis by producing 'primarily for corporate profits.' They called for an intensive search for 'alternatives to the prevailing Western Economic System.' "[6]

2. They have had particular appeal to college students and young persons. Those young persons soon will be in important decision-making positions, and they are already affecting the powerful decision makers of today. Some of the most talented college graduates, often the ones businessmen most want to employ, are demanding to know what the corporations are doing for the community and for the nation.

The young persons are pressing their elders to initiate business reforms. The head of the National Urban League was reported as startled "to see the chairman of one of the largest U.S. corporations moved to tears while describing the persistent questions of his children as to what he, as a man of power, was doing to improve society."[7]

3. Helped along by such developments as unpopular wars, the growing resentment of the military-industrial complex, the revelations of public and private corruption, repeated instances of widely publicized violations of the anti-trust laws, frightening inflation attributable in part to the misdoings of giant corporations, and the problems of nuclear power, the New Left has created a general climate of distrust of the economic system.

4. The reasoned and scholarly attacks of the New Left have been given some credence by a large group of middle-class reformers. The author of *The Radical Attack on Business* concludes:

> It is quite likely that the New Left has been partly responsible for the renewed activity of liberals since the 1950's . . . and has helped to legitimize and fuel the middle-class insurgency of Ralph Nader, Common Cause, consumerism, ecology, cultural experimentation, and anti-war movements.[8]

Reformers and Fellow Travelers

The Author's "Aside"

That which follows is intended as a sympathetic but objective analysis of the contemporary attack on business. We believe:

1. there are social and economic evils that need correction;
2. society's ills will never be corrected unless some persons are sufficiently dedicated to serve as leaders;
3. as human beings, leaders have weaknesses, and, often enough, their weaknesses are our weaknesses.

The Chief "Causes"

The chief ideological "causes" today are: (1) environmentalism, (2) consumerism, (3) world peace, and (4) humanitarianism. All of them have substantial impact upon the business system. It may well be that none of them can make great progress except at the expense of at least one aspect of the business system.

The Exponents

The exponents and supporters of each of the causes can be divided into the following categories:

1. The leaders. The leaders can generally be described as: inspired, dedicated, unselfish, intelligent, and able. Almost invariably, they share the following weaknesses.
   a. Wrapped up in their particular cause, they tend to be simplistic. They are prone to overlook the complexity of society, the multiplicity of variables, and the difficulty of distinguishing between cause and effect.
   b. They frequently overlook or ignore the possible adverse consequences of their own activities. They usually do not carefully weigh the gains against the losses. They spend more time in attacking "evils" than in seriously seeking viable alternatives.
   c. They welcome assistance from any source that they deem helpful to their cause. They ignore, or make little effort to ascertain, the undesirable characteristics of hangers-on. They seldom denounce irresponsible charges made by followers in the name of the cause, charges which they would never make themselves.
2. The "Fully-Persuaded"
   Among the full-persuaded are many able persons who are as sincere as the leaders, but simply are not so fully committed. Among them are most often found: journalists, religious workers, social workers, college professors, idealistic students, and some high-minded politicians. The list is not all-inclusive. Depending, perhaps, on the extent of their commitment, they share to some degree the strengths and weaknesses of the chief leaders.
3. Fellow Travelers. The possible fellow travelers are too numerous to list in their entirety. Some of them are:
   a. revolutionaries seeking the destruction of the social order and ready to attack at any point seeming to be vulnerable,
   b. the greedy, looking for the chance to make a fast buck,
   c. the insecure and inferior, looking for publicity,
   d. the lunatic fringe, enjoying thrashing about,
   e. the alarmed, willing to try anything, and
   f. the gullible, simply following those who say they have "truth."

The Attack in 1976

Whether the attack of 1976 should be labeled "attack on business" or "reform movement" probably depends upon one's point of view. The extent and pervasiveness of the movement as well as examples of some of the personnel just described can be seen in three newspaper article appearing on the same day, 22 April 1976, in the *Toledo Blade*. The headings are our labels of causes or leaders. The newspaper's headlines or leads are in quotes.

1. *Young Revolutionary*
       "Backers of Revolutionary Concepts Stir Rebellion by Some in Business"
   The article, briefly paraphrased, occupied almost two newspaper columns. Mr. Jeremy Rifkin, founder and president of the Peoples Bicentennial Commission, spoke at the University of Toledo. His appearance was sponsored by the Student Union Board.

The Peoples Bicentennial Commission claims 2,000 members and advocates and predicts a full-scale revolt to replace the current economic system.

2. *Consumerism*
"Consumer Impact News Promised"

Mrs. Joan Braden, Consumer Affairs Coordinator at the State Department, at a salary of $37,000 per year, spoke to about 40 persons at the annual meeting of the Consumer Council of World Trade. She is one of seventeen consumer affairs coordinators appointed by President Ford in December 1975. Their primary function is to make federal agencies more responsible to American consumers.

3. *Environmentalism*
"Environmental Watchdog, a Ph.D. Politician!"

Dr. Peterson is a chemist who helped develop Dacron for Du Pont and later served as Governor of Delaware. In 1976, he was chairman of the President's Council on Environmental Quality. He came to Toledo to speak to the Toledo Section of the American Chemical Society.

Dr. Peterson "won national fame for his fight to save Delaware's beautiful Atlantic Coast and Delaware Bay beaches from further encroachment by heavy polluting industries." The financial costs of his reforms forced him to sponsor unpopular tax increases and contributed to his defeat for reelection as governor.

Two other items are offered as also representative of the visible signs of the times. The first appeared in the *Unitarian Universalist World,* 15 April 1976, just one week before the articles noted above.

1. *Church and Stockholder*
"Corporate Responsibility Is Goal of UUA Effort"

The UUA (Unitarian Universalist Association) announced success in cooperation with "The Interfaith Center of Corporate Responsibility." That center "is a group related to the National Council of Churches." The announced success was an agreement with Avon Products to "make public that company's policies and programs concerning equal employment opportunity."

The Reverend Leslie W. Cronin also announced that the Center had filed stockholder resolutions with American Home Products and Colgate Palmolive. From American Home Products, the Center sought "information about the sales of infant formulas in developing countries, including a list of all foreign countries in which the company markets a formula, health promotion policies, nutritional information and research."

2. *Environmentalism*

Also in 1976, California held a statewide referendum on a proposal to "safeguard nuclear power installations." As was well known, the referendum proposal was a thinly disguised effort to completely eliminate nuclear power installations. The referendum lost at the polls, but almost certainly because the issue was defused by the State legislature. At the eleventh hour, the legislature passed a law which provided more moderate regulatory changes. That the pressure on business for social responsibility is increasing is recognized by all, including business leaders.

The Social Audit

The August 1973 edition of *Recreation Management* carries excerpts from an address made by Donald E. Garretson as vice president of Minnesota Mining and Manufacturing Company. The first three paragraphs of those excerpts are quoted:

> As a CPA and a financial officer of a large company, I know a thing or two about audits. Financial audits serve a very useful purpose in business, as I need not remind you.
>
> We hear more and more about a new kind of audit, called a Social Audit . . . and it, too, can serve a very useful purpose in business. As a financial officer, I am as concerned about the results of the Social Audit as I am about the purely financial audits of my company's operations. And you, as recreation directors, should be, too.
>
> If business, generally, flunks on its social audits, then business and society will become adversaries in a dispute which neither can win and in which all of us will certainly be losers [p. 18].

The Climate of Opinion

Apparently between the time of Garretson's address in 1973 and August 1976, business in general had flunked its social audit, and the climate of opinion became quite unfavorable to business.

Not withstanding that many individuals believe they are quite objective and base their conclusions on "the simple facts of the case," the climate of opinion affects everyone: legislators, judges, journalists, and the man on the street. The consequences for American business include:

restrictive legislation   increased regulation
adverse judicial decisions   opposition to needed rate increases
negative zoning decisions   heightened labor conflict
multiplication of civil suits   employee disloyalty
greater difficulty of recruitment,
especially of college graduates.

Illustrating one cause and several results of the unfavorable climate of opinion are five items appearing on the same day, 3 August 1976, on the front page of the *Wall Street Journal.*

1. "Foreign payoffs were disclosed by two more companies, Squibb and City Investing."
2. "Anti-trust bills making it easier for government to block big corporate mergers and strengthening the Justice Department's antitrust investigatory powers won House approval."
3. "Prescription-drug sales practices investigated by the Federal Trade Commission to see if they or private actions are depriving consumers of adequate price competition and product information."
4. "Grain-inspection has spread to at least fifteen states. . . . "
5. "Compensation Rift: Labor and Employers Clash over Changes in Disability Payments. Both Say Change is Needed: Michigan Issues Illustrate Controversies across U.S. Fears of a Business Exodus."

The number of such items in one issue of one newspaper may have been somewhat exceptional, but on other days in other newspapers and magazines, similar articles appeared with almost monotonous regularity.

The Trial of Capitalism

Wherever one looks, there are the stresses and strains of capitalism: the confrontation of labor and management, the antithetical interests of industrial growth and environmental sanity, the restless thrashing of minorities, the antagonism of the underdeveloped nations of the world, and the drift toward socialism. Capitalism does appear to be on trial. It was a serious and not a redundant question that was asked by the cover story of *Time* on 14 July 1975: "Can Capitalism Survive?" [pp. 50–63].

In every democratic, capitalistic country of the world we can observe the uneasy shifting of the balance of power between and among governmental regulation, massive unions, and giant corporations. It is not comfortable to realize that if the scales tip too far in either direction, the consequences can be disastrous.

There are many persons who question whether capitalism, even with the resiliency it has shown thus far, can always manage to adjust rapidly enough to accomodate the stresses and strains that seem to be building up with increased vigor and frequency. Economist Milton Friedman "darkly suspects that capitalist freedom will turn out to be 'an accident' in the long sweep of history."[9]

The Story of Capitalism

Historically capitalism is in its infancy. In the same year that the American colonists signed the Declaration of Independence, Adam Smith published *An Inquiry into the Nature and Cause of the Wealth of Nations.* That book, commonly referred to as *The Wealth of Nations,* is regarded as marking the real beginning of the capitalist system. Obviously full-blown capitalism did not spring into being over night. Its rudimentary beginnings may be seen much earlier. But Adam Smith's book made it visible as a system—put flesh on the skeleton, so to speak. Also 1776 was in the dawn of the so-called Industrial Revolution. The spinning jenny had just been invented (1764), and the power loom (1785), and the cotton gin (1793) were soon to come. Adam Smith's book truly ushered in a new age.

Man has been on this earth something like two million years; he began farming, the first well organized economic activity, about ten thousand years ago; and learned to write, ushering in the historic age, about five thousand years ago. In this perspective, the roughly two hundred years of capitalism is indeed a brief period.

Adam Smith

Adam Smith saw "natural" social forces developing into a system [Karl Marx gave that system the name of capitalism] which Smith believed would enthrone the consumer, not the businessman. His basic premise was that the competition created, quite unintentionally, by business operations would fuel an economic system, a self-regulating mechanism, serving the best interest of society. Smith believed that the profit motive would cause entrepreneurs, each acting in his own "enlightened self-interest," to compete with each other in such a fashion that the free marketplace would produce the commodities which society most desired in the quantities needed and at prices which society was willing and able to pay.

Monopoly

Adam Smith did not foresee the power of monopoly as wielded by industrial giants. Carnegie, Rockerfeller, and the other so-called "Robber Barons" proved that in a free market they could operate in such a fashion as to create tremendous wealth, and power, and then more wealth. Ultimately they were able to achieve monopolies. Monopolies provided the power to manipulate markets, prices, and supplies, at the expense of the consuming public and of the workers who produced the goods.

Regulation

The evils of monopoly persuaded the United States, over a half century ago, to embark upon a deliberate policy of legal and judicial actions to regulate the railroads and other industries. Particular emphasis was placed, originally, upon the prevention of "combinations in restraint of trade." From the beginning—and there is no end in sight—there has been the steady growth of legislation, regulation and regulatory bureaucracies, and of the areas and aspects to be regulated. Each increase in regulation more narrowly circumscribes the freedom of business and industry, *and the consumer*! Capitalism also has other problems.

Other Problems of Capitalism

Inflation-Depression

One serious question is whether, in a democratic and free society, the economy can be so managed that inflation will not eventually destroy the system. Many persons are concerned about the inflation-depression cycle which, in recent years, has upset the confidence once engendered by the theory of John Maynard Keynes. Keynes and his disciples believed that an informed government could permanently eliminate dangerously wide swings of the economy. Theoretically, through such devices as taxation, public spending, and currency control, the government could so stimulate or restrain economic growth as to control either inflation or recession. While government undoubtedly can exert great influence on the capitalistic system, the realities of practical politics has created serious doubts as to whether the government can cure recession without engendering inflation which will bring about a new recession. As it happens, politicians at all levels pay more attention to the clamors of their constituents, the voters, than they do to economic theorists, who are never in agreement anyway!

Inequality

A deeper problem of capitalism is rooted in morality and ethics. Could economic prosperity have come about to the extent that it has without feeding upon the inequality of women, children, Blacks, and other minority groups? In the past, the Blacks, for instance, have accepted a very unequal share of the rewards of capitalism. They seem disinclined to continue to do so.

It is true that, in general, individuals are better off and have more worldly goods than did their parents and grandparents before them; but their expectations have risen more rapidly than their wealth. This is partly because people are better educated and more aware of what others have and of what is available. It is also true that the more people have, the more they want.

Political Equality

Accompanying greater economic democracy has been greater political democracy. Women's right to vote in all elections is little more than a half-century old. The actual, practical ability of Blacks and other minority groups to use their legal right to vote has become nationwide only during the last dozen years. The right of persons 18 to 21 years of age to vote in presidential elections was first exercised in 1972. All these groups are important segments of the labor force.

People use their votes to elect candidates who promise to legislate the benefits that the voters want. Does the increase of political democracy mean the multiplication of laws, regulations, and judicial decisions all designed to *force* economic equality?

Liberty and Equality

An unresolved question is just how much inequality will be and can be tolerated within a system which can still be called democratic capitalism. It is unfortunately true that liberty and equality are antithetical concepts. We cannot have more of either without less of the other. We cannot eliminate equality without eliminating freedom.

We should remember that, theoretically at least, managed economies can produce as much real prosperity as free and democratic capitalism. However, that has yet to be demonstrated over a long period of time in actual practice. We do know that, whether the managed economy is as far to the left as communism or as far to the right as absolute dictatorship, any gain in equality is at the price of freedom.

Marx and Engels

Karl Marx and Friedrich Engels, his close associate, believed and predicted that capitalism would ultimately destroy itself. They lived later in time and had the advantage of witnessing developments that Adam Smith did not foresee. The *Communist Manifesto* was proclaimed in 1848, over 70 years later than the *Wealth of Nations*. The more completely developed theories of Marx and Engles were published in *Das Kapital,* in three volumes appearing between 1867 and 1894. The last two volumes were written by Engels, mostly from the notes of Marx, after the death of Marx. Thus, *Das Kapital* came roughly a century later than the *Wealth of Nations.*

Marx and Engels were able to observe that the natural forces of competition did not result in a free marketplace and enthrone the consumer. They saw the accumulation of wealth and they observed that wealth is power. They witnessed the rich getting richer and the poor becoming poorer. Small wonder they predicted an ultimate class war!

Marx and Engels were in error in some respects because of developments which they did not foresee. Among those are universal education and universal suffrage. Nevertheless, we cannot say yet that they were wrong in their basic prediction that capitalism will ultimately destroy itself. Its fate still hangs very much in the balance. To state today that Marx has been disproved in his basic prediction is no more logical than is the old man who thinks he never will die because he never has yet.

Smith and Marx Contrasted

Smith based his theories on the premise that "natural forces"—men seeking their own immediate satisfaction—would create a free marketplace to the benefit of society. The end result would be accidental but fortuitous. Monopolists proved him wrong in that basic premise. The addition of government regulation has not proved a solution and offers no real promise of a solution.

Marx based his prediction on the principle of "natural" conflict and struggle. Out of that conflice, Marx envisioned the victory of the common man. If we have learned anything at all in the years since they wrote, it is that in wars there are no real winners, only losers. That is true whether the war is military or economic. In the long run, an adversarial relationship is harmful to all participants.

A Ray of Hope

Without attempting any predictions, we do see a glimmer of hope for our time and our children's time. That hope lies in the beginning of a new approach, both in theory and in practice. Perhaps through experience and with the help of universal education we have gained a small amount of wisdom; and perhaps through universal suffrage we can put that wisdom to work for all of us.

Since we have lost our faith in the "natural" forces of competition and conflict, some people are turning to deliberate planned approaches. It does not mean the abandonment of the profit motive but the tempering of it with self-restraint and consideration for our fellow man. It means the embracement of social conscience.

The Social Conscience of Business and Industry

It is not really that the idea of social conscience or social responsibility on the part of business and industry originated only yesterday; it is just that the concept was accepted and practiced by too few. Prefacing this chapter is a statement by Samuel C. Johnson taken from *Top Management Speaks*. In that statement, Mr. Johnson, Chairman of the Johnson Wax Company, quotes from a 1927 address of his grandfather:

> When all is said and done this business is nothing but a symbol. . . . In a very short time these lively machines will become obsolete and the buildings for all their solidity must some day be replaced. The goodwill of the people is the only enduring thing in any business. It is the sole substance and the rest is shadow!

It was the critics of business who first said that business must develop a social conscience, and a few years ago many people considered that sentimental poppycock. Today it is being said by businessmen themselves and more persons are taking it seriously.

The Business Trend

In recent years there is a growing trend for business and industry to use self-restraint and to act as though a social conscience is being developed in the commercial sector. The trend was small at first, a tiny stream, but there are indications that it may grow into a broad river. There is stubborn refusal to consider the concept by some businessmen, and there is ambivalence on the part of others; yet, in general, the concept seems gradually to be gaining acceptance. We illustrate the trend by a number of published articles in years between 1968 and 1976. All of them are from business-oriented periodicals, most of them from the business organs of the Conference Board. Titles are in quotes and dated by years.

"The Role of Business in Public Affairs" (1968)

In 1968 the Conference Board published a research report on "the role of business in public affairs." In the Foreword, the editor expresses some optimism.

> It is encouraging to note, for example, that three out of every four responding companies now have a recognizable public affairs function—compared with a mere handful of firms only a decade ago.
> Especially important is the indication that management is taking a much broader and more penetrating look at socio-economic problems than in the past, and is expressing a willingness to initiate action to help solve them.[10]

The study report did note conflicting views. One company representative asserted: "Our business is to manufacture and sell our products, not to solve the problems of the nation." On the other hand, one manufacturer responded: "The company will maintain an active interest in the general welfare of our society. . . . The company regards itself as a citizen of the community in which it maintains facilities and employs people."[11]

"The War That Business Must Win" (1969)

This article extolled the social activities of some companies and cited some specific examples, yet it noted the unevenness of acceptance of the concept of social responsibility. "Some companies have attacked social problems boldly, others have flagged. One major lesson: much remains to be done."[12]

The article noted that public service projects were more likely to fail where companies were "responding purely out of some lofty sense of mission" and seemed to enjoy the greatest success where enlightened company leaders responded to the profit motive. In that context it quoted an important statement by Henry Ford II. Though brief, we set it off for deserved emphasis.

> Improving the quality of society—investing in better employees and customers for tomorrow—is nothing more than another step in the evolutionary process of taking a more far-sighted view of return on investment.[13]

"The Executive as a Social Activist" (1970)

The writer of the article notes that the actions of business leaders does not always match their rhetoric, but concludes that some progress is being made. Possibly the most significant change was that "more and more businessmen are becoming aware that social action is not a luxury; they realize that either they have to change or that it will be forced upon them."[14]

"America's Social Crisis: Future Perspectives" (1971)

Perhaps the most significant aspect in 1971 was that the Conference Board, representing business interests as it does, should have held a special conference on "Business Leadership in Social Change." The article cited above is from an address made at that conference. The speaker was quite realistic and noted both negative and positive aspects. He, too, felt that the most promising steps were those taken by business on the basis of the profit motive. He listed some such projects and commented: "While these programs may also be in their sponsors' interest, they promise real contributions to the solution of some of our social problems. They are the forerunners, I think, of more massive and organized business efforts over the next ten years."[15]

"Of Anniversaries and Revolutions" (1973)

In an inspiring address, John D. Rockerfeller III suggested that America is in the process of a second revolution. The first revolution, he observed, was daring and "presented a new . . . vision of a just and humane society." Realistically he noted that during the past two hundred years we have sometimes honored our ideals and at other times "ignored them shamefully."

His conclusion was that "the central task of the Second American Revolution is to reconcile our dominant materialistic values with our enduring humanistic values—to preserve the benefits of economic and technical progress, but to order them within the context of a society which is increasingly humanistic and person-centered."[16]

"On the Mutuality of Our Interests" (1974)

James F. Bere expressed concern that the public is increasingly reacting emotionally and negatively to the activities of business. For the situation, he blamed businessmen themselves because "as for what business is doing wrong in terms of its image today, it is hard to pick up a paper without finding some glaring evidence. And I don't mean unintentional mistakes, but rather deliberate and repeated errors of judgment. . . .

If we are to solve the problem of our public image, he concluded, "we must first persuade *ourselves* of the mutuality of interest between businesses large and small, labor unions, housewives, consumer organizations, farms, governments, and other groups."[17]

"Two Roads to Social Responsibility" (1975)

Walter A. Hamilton refers to the large number of rules and regulations forced upon business but expresses the opinion that the rules and regulations are "the legal embodiment, the codification of the judgment of our society, as to our corporate social responsibilities."[18]

The "two roads" of which he speaks are (1) voluntary and constructive approaches to social responsibility, or (2) forced compliance with the rules and regulations. Of the second course, however, he warns that "the adverse impact of such a posture upon public attitude toward them can be costly indeed when the time comes for labor negotiation, zoning approvals, or consumer choice in the market place."[19]

"Productivity and Employment Security" (1976)

Bruce Thrasher, a labor union leader explains that union leaders are finding that labor's objectives can often be better achieved through cooperation than through conflict. The entire article is impressive, but most significant is his statement that the United Steelworkers of America believe "that a stable and profitable steel industry is of vital importance to the economy of our nation, and that it is also important to the steel companies themselves and to their thousands of employees who are members of our union."[20]

The Role of Industrial Recreation

We have tried to point out in this chapter that capitalism has really grave problems, the outcome of which are still in doubt, but that there may be an answer in abandoning conflict in favor of cooperation. We do not wish to exaggerate the importance of industrial recreation to the solution of the serious problems of captalism, but we sincerely believe that industrial recreation can make a significant contribution. Industrial recreation exemplifies mutuality of interest. It is, as Mel Byers, CIRA, emphasizes, a mutual benefit association, of mutual benefit to the employer and to the employees. It also can be, and in a growing number of instances is, of mutual benefit to the community.

In this respect we direct attention to one sentence in the statement of Paul P. Davis which prefaces this chapter.

> No longer do we consider recreation a fringe benefit, it is the catalyst that produces healthy, vigorous, dedicated employees who are an asset to McLean [Trucking Company] and outstanding citizens in their communities.

We close the chapter with a long quotation from the July 1975 issue of "Keynotes," published by NIRA, and authored by our mentor, Mel Byers. In respect to mutuality, we cannot conceive of anything more relevant to this chapter.

> Industrial recreation and employee services are one of a few remaining management prerogatives. They are the unnegotiated benefits that set one company ahead of another. Properly administered, there is no equal or even a close comparison to reaching the employee and developing a cooperative, understanding attitude.
>
> The efforts put forth to build harmony and consideration are far more profitable than that of battle line strategies and war maneuvering. The successful history of companies providing excellent employee service programs prove the values through fewer confrontations between employee and employer, more pleasant and harmonious relationships, and a greater employee interest and concern for the product and the company.

Not by coincidence, of course, the passage from "Keynotes" leads us naturally to Chapter 3, which is about the company benefits of industrial recreation.

Notes

1. "Can Capitalism Survive?" *Time.* 14 July 1975, p. 63.
2. Charles Perrow. *The Radical Attack on Business: A Critical Analysis.* (New York: Harcourt, Brace, Jovanovich, 1972), p. 9.
3. Ibid., pp. 10–11.
4. Ibid., p. 1.
5. "Can Capitalism Survive?" *Time.* 14 July 1975, p. 54.
6. Ibid., p. 50.
7. "The Executive as a Social Activist," *Time.* 20 July 1970, p. 62.
8. Perrow. *The Radical Attack.* p. 288.
9. "Can Capitalism Survive?" p. 50.
10. H. Bruce Palmer. "Studies in Public Affairs, No. 2." A Research Report of the Conference Board, p. iii.
11. Ibid., p. 25
12. *Business Week.* 1 November 1969, p. 63.
13. Ibid., p. 64
14. *Time.* 20 July 1970, p. 63.
15. Theodore J. Gordon. *Conference Record.* July 1971, p. 35.
16. *The Conference Board Record,* July 1973, p. 13.
17. Ibid., July 1974, pp. 48–49.
18. Ibid., July 1975, p. 20.
19. Ibid.
20. Ibid., July 1976, p. 22.

3

Company Benefits of Industrial Recreation

TOP MANAGEMENT SPEAKS

The purpose of an industrial recreation program, of course, is to foster a sense of belonging . . . a sense of individual and company pride. Years ago, this was accomplished primarily with company baseball and softball leagues, employee bowling teams, and other competitive activities. But just as industry has changed, so have industrial employees. Today, their interests continue to include team sports, but they also encompass a broader range of opportunities for self-expression.

At United States Steel, our employee organizations—one of which was begun almost seventy years ago—put together diverse programs of year-round activities designed to meet the recreational interests of almost every individual. In addition to bowling, golf, tennis, and other sports, there are opportunities to enjoy local cultural events, to travel in groups to vacation spots and foreign lands, to socialize at off-hour dances and picnics, and most important, to share one another's interests in music and hobbies.

A very special recreational outlet for our people is a Christmas Choral Festival held annually at U.S. Steel's headquarters in Pittsburgh, Pennsylvania. It is planned, coordinated, staged, and directed entirely by U.S. Steel employees. Thousands of their co-workers and their families attend, and since the event is open to the public, they are joined by many other thousands of local citizens.

For three days, there is musical entertainment provided by choral and orchestral

Edgar B. Speer

groups from our various plants and offices. Simultaneously, there is figure skating by sons and daughters of our people. There are displays of employee talents in the arts and crafts. Hundreds of women from our headquarters office dress dolls in a variety of costumes that are displayed to the public, then distributed to needy children.

A year ago, following an evening performance by all of the musical groups, one of our employees was overheard remarking to another, "It makes you proud to be a part of U.S. Steel, doesn't it?" We've found that sentiment to be a common one at Festival-time. And when you come right down to it, that's really what industrial recreation is all about.

Edgar B. Speer
Chairman
United States Steel Corporation

TOP MANAGEMENT SPEAKS

A balanced recreation program offers many benefits beyond its admitted contributions to physical and mental well-being. Over the years, we have noticed that employees who participate in the many and varied activities of our 3M Club tend to remain with the company.

The personal friendships formed with fellow employees from all areas of the company contribute to a feeling of belonging and help promote the small company atmosphere that could diminish as a firm grows. This feeling of belonging and wider acquaintance within the company helps the employee bring a little something extra to his job. That is important both to him and to 3M, where we have a long standing policy of developing our own leadership and promoting from within.

The highly creative, talented people who contribute so much to the progress of our company are interested in a multitude of activities. The successful recreation program is one that is tuned to this variety of interests by all empoyees—young and old, men and women, the athletic and not-so-fit. Still, I'm sometimes a bit surprised at the spectrum of leisure time pursuits that interest our people: from music

Raymond Herzog

to chess, foreign languages to calorie counting, photography to philately, as well as sports for every season.

Our emphasis, as it has since the beginning of our recreation efforts, continues to be on family participation. We feel that the 3M program exerts a strong, positive influence on the quality of employees we attract and retain. We look at it as part of our investment in the human resources that are the key to 3M success.

Raymond Herzog
Chairman of the Board and Chief
Executive Officer
Minnesota Mining and Manufacturing
Company

Industrial recreation programs are of great benefit to companies through increased production, which can be translated into profits and growth. Nevertheless, just in the United States, there are thousands of companies, some quite large, which do not have organized industrial recreation programs. There are even more which have programs which are inadequately supported and not provided with professional leadership.

Increased Productivity

The reason that there are not more adequately organized and well funded industrial recreation programs is that concerned employees are reluctant to go to management with strong request for company support for recreation programs because they do not know how to deal with the objection they expect to get. If they go, the management response often is: "This is a business organization. The company can afford to support only those activities which pay for themselves and return a profit. If you can show me that a recreation program will pay for itself—show me that it increases productivity—the company will provide the support."

Management cannot really be blamed. The president of a company has to be concerned about the bottom line of the balance sheet; that is what he is paid for. Basically his board of directors have only two questions for him. They are: How much money did the company make this year? and What is the outlook for the future? Except for a tiny though perhaps growing, percent of the company's stockholders, those are the only question they ask.

We should not confuse power with security. Except for the company which is family owned or controlled—less and less often the case today—the corporate president who gives the wrong answer to those two questions today may be hunting a new job tomorrow.

Neither can we blame the employees who do not go with the request or are unable to deal adequately with the usual objection. The truth of the matter is that no one has ever been able— and very likely no one ever will be able—to prove that industrial recreation increases productivity. A few persons have made feeble attempts; and some have come up with answers that satisfied themselves but few others. The reason is that productivity is the consequence of so many variables that it is verging upon the impossible to develop and fund a research project sufficiently large and sophisticated that it can control all the variables and arrive at a scientifically defensible conclusion.

On the other hand, if you can demonstrate that industrial recreation can improve communication within the company; discover and develop leaders for the company; improve employee morale and physical fitness; recruit and retain better quality employees; reduce absenteeism; and prepare employees for retirement—if you can do all that, who will ask, "Does it improve productivity?" Those are our objectives in this chapter.

Communication

Communication is one of the most important and difficult aspects of organizational management. It would be hard to exaggerate either its importance or difficulty.

The means of communication are so limited in general use: words, which probably never mean exactly the same thing to sender and receiver; and behavior, or looks, or pictures, all as frequently mininterpreted. Messages are colored—or discolored—by our emotions, attitudes, preconceptions, by our likes and dislikes, and even by our feeling toward the media, human or otherwise, through which the messages are transmitted.

We hear (accept, believe, pay attention to) only those messages which are:

a. transmitted when we are in a favorable mood,
b. about things which we wish to believe,
c. sent from suitable sources (whom we like, trust, respect),
d. transmitted through a suitable medium (a person or thing about whom or which we have good feelings), and
e. in a form we can easily understand.

There is no communication unless the set is turned on and only very poor communication unless the set is properly tuned in.

Most of the time, communication is not at its best unless it is two-way. There needs to be feedback: agreement, disagreement, questions, indications of comprehension, or acceptance or rejection—all so that the signal can be continued, repeated, or modified as often as needed to insure that sender and receiver are in tune and share an idea.

Need for Communication

Many of the difficulties that exist in business arise from the fact that the rank and file do not know what is really intended by management. Signals from management are weak, contradictory, or nonexistent.

Management more often does not receive information from below. There is a lack of valuable constructive criticism—not infrequently the employees know better than anyone else what is wrong with management, where the problems are, and why plans go awry. Creative or problem-solving ideas of line employees do not float upwards and receive attention. Lacking information about workers' frustrations, management can never remove them.

In 1964 the Conference Board conducted an extensive study through industry questionnaires of the extent and media of intracompany communication. In the Foreword of the report, published in 1966 as "Employee Communication: Policy and Tools," H. Bruce Palmer emphasized the importance of good communication: "The Conference Board has long identified good communication as essential to good management, and over a period of 25 years has conducted studies in depth of various media."[1]

Downward Communication

Downward communication is almost everywhere recognized as extremely important. Its many benefits are illustrated by the remarks of respondents included in the Conference Board report. Five of the many are included here.

1. Efficient work performance cannot be expected unless an employee knows how his work is to be done and how it relates to the functions of the entire plant or office in which he is employed.
2. We believe that productivity is influenced by how well an employee understands, accepts, and supports the actions that management want taken.
3. Informed employees will be better interpreters of their companies in outside contacts.
4. Informed employees who understand the "why" of their jobs as well as "what" will produce better and help keep costs down.
5. [The purpose of communication is] to develop favorable attitudes and support of the company by employees. . . . Employee opinion is often adversely formed by misinformation or even as a result of no information.

Upward Communication

Many persons tend to think of company communication only as what the boss tells the workers; and, indeed, that is almost the whole thrust of the 1966 Conference Board Report, No. 200. Now and then we do see a glimpse of more understanding. Probably the best illustration was the statement quoted from Lynn Townsend.

> Every member of management must understand that effective communication is an essential tool of good management; and that part of his job is to relay appropriate information and news, whether good or bad to his subordinates and superiors. . . .
>
> Most employees basically want to help a business get better—and this is very significant. But we've got to talk and listen to them. And we've got to do it regularly, not on an on-and-off basis. . . . To unblock upward communication channels is to tap a great reservoir of creative opinions and suggestions which can be of great help in obtaining the legitimate goals of the corporation.[2]

The Conference Board report included a summary table devised to reveal the extent to which various kinds of media were used. The indicator was the percent of respondents using each of the media. The table lists 26 different media employed. Of those, only three were specifically designed for the upward flow of communication.

Of those designed for upward communication, the only one showing substantial use—by even half of the responding companies—was the exit interview. Exit interviews are undoubtedly useful but they get information from the employee about his or her feelings and attitudes only after the employee is forever lost to the enterprise.

At the time of the survey (1964), only 20% of the respondents made use of attitude surveys, one of the other two listed media for upward communication. By 1974 employee attitude surveys were the "in thing," and many companies were using them. The *Wall Street Journal* of 26 March 1974, introduced an inside story by the front page item:

> Employee surveys grow more popular as firms probe workers' attitudes.
>
> . . . Responding to worker gripes discovered in surveys, General Motors recently revised the management system of a Georgia car plant.
>
> . . More unions are cooperating with surveys a Chicago consultant says, because *"even the unions can't figure out what's on the* worker's mind" [emphasis added].

Apparently little if any change had occurred by 31 March 1976, the date of an author interview with John H. Rapparlie, Ph.D., industrial psychologist for Owens-Illinois, who frequently serves as an outside consultant for other companies. "Outside morale studies are necessary," he observed, "because management never seems to know what employees are thinking." He added, "I have never found a situation where an employee group has said, 'We don't need a survey.' "

Dr. Rapparlie observed that, as one technique, "I always ask management ahead of time what they can see as their major personnel problems, and management can almost never foresee the results. Management simply doesn't know what employees are thinking."

Asked if he could generalize on the most common errors of opinion by management, Dr. Rapparlie responded, "Management is more likely to see problems in the area of pay and fringe benefits such as the amount of vacation time; whereas the most frequent complaints of employees are inefficient plant operation and the disregard for employees' ideas."

Dr. Rapparlie's views of management's ignorance of employee attitudes is borne out by the above story from the *Wall Street Journal* and by other sources. Harold Mayfield, then corporate director of personnel for Owens-Illinois, wrote in his "Personnel Newsnotes" of April 1962: "One of the first scientific studies of employee attitudes revealed that managers are not good judges of what people want from business—and more startlingly that top labor leaders are not good judges either."

Lateral Communication

Though management is finally recognizing the need for upward communication, the need for lateral communication apparently has still received very little attention. In the 1966 report of the survey study done by the Conference Board, we did not discover any mention of lateral communication, and certainly none of the 26 media listed in the survey table of results was designed specifically for that purpose.

In company after company, in a figurative sense, the left hand does not know what the right hand is doing. And quite literally, different departments of the same company are frequently working at cross purposes. In many companies, important developments in one department are unheard of by a department down the hall. The consequent departmental isolation causes confusion, contradictions, and clashes between employees. It is an important contributor to company mistakes and inefficiency.

There is a very important and relevant statement in *Top Management Speaks* by Sherwood L. Fawcett, President of the Battelle Memorial Institute: "We believe that a well-developed program of recreational activities . . . stimulates the exchange of ideas—a particularly important element in interdisciplinary research [preface to chapter 1]."

Communication via Industrial Recreation

Industrial recreation was not taken into account as a communication medium in the Conference Board report of 1966. But then, it was based upon a survey of 1964, and industrial recreation programs have grown tremendously since that time. If a similar survey were to be undertaken today, it is inconceivable that recreation programs should not receive attention.

Improved communication throughout the breath of a company is one of the more important benefits which the company derives from recreation programs. During the course of recreational activities, the meeting of special clubs, and the performance of its various kinds of services—all of the time the recreation program provides communication par excellence.

Workers and supervisors up and down the chain of command come together informally and learn to know and trust each other. Workers of different departments know each other as they never would otherwise, and information flows between them during recreation programs. More than that, since they know each other, each knows someone in another department with which contact is desired in the performance of their official work activities.

Some of the people in the recreation program, and certainly the director, ought to be in direct contact with top management. If there is a good industrial recreation program, there is no reason at all why management should not at all times know what the employees are thinking.

The employees come to know their supervisors well and also know other representatives of management. During the recreational activities there is no reason to restrict communication to official channels. Upward and lateral communication is made easy.

Management which is familiar with industrial recreation knows the programs' usefulness in communication. Robert W. Galvin, whose statement prefaces this chapter is one of the more enlightened executives, and communication is one of the things he emphasizes:

Motorola's recreational activities are a natural, mutually enjoyable extension of wholesome, on-the-job relationships. Our program has long been one of the strong links in the communication chain among us all.

No other communication medium possesses all the advantages of industrial recreation organizations. They—

a. are the means by which messages can be transmitted under favorable, emotional, and attitudinal circumstances,
b. provide media—intermediates—who are known and trusted,
c. permit of two-way communication, and,
d. are equally effective for downward communication, upward communication, and lateral communication.

These are the benefits that can come from industrial recreation, but the extent to which the benefits are realized will depend to some extent upon whether communication is a specific objective of the program and of the company.

Leadership

The identification, testing, and development of leadership through industrial recreation programs is another illustration of the mutual benefit aspect of industrial recreation—the company benefits through the acquisition of a good supervisor and the employee benefits through the opportunity for promotion.

There is no doubt at all that the identification of potential leaders is a substantial company benefit. *The Wall Street Journal* of 27 August 1974 reported that "managerial talent is scarce and will be getting scarcer." The article concerns a study done by Professor John B. Miner, of Georgia State University for the Bureau of National Affairs. To ease the shortage, Professor Miner suggested "better use of women and minorities and improved corporate searching for potential managers."

In doing a research study of the employees of the Johnson Wax Company for another purpose, Reginald Carter, Ph.D., observed that "representatives of both management and hourly paid employees are frequently involved in the same sport or activity [and that] in some cases such opportunities for leadership in organizing these activities has resulted in the discovery of latent supervisory skills among both blue and white collar personnel."

In the instance of the study at the Johnson Wax Company, Dr. Carter considered the discovery of leadership qualities to be sheer good fortune. However, the discovery of potential leadership occurs more regularly and more frequently in programs which have the discovery and development of leadership as one of their deliberate objectives.

To some extent, also, the extent of leadership discovery and developments depends partly on the attitude of the company, though there must be few who do not welcome such an outcome. Raymond Herzog, chairman of the board and chief executive officer of Minnesota Mining and Manufacturing Company, notes in the preface to this chapter that at his company, "We have a long standing policy of developing our own leadership and promoting from within."

John H. Rapparlie, Ph.D., corporate industrial psychologist for Owens-Illinois, stated in an interview with the author that large companies spend millions of dollars on training management "almost solely on faith." He expressed the belief that there is no test and no interview system which can be right a large enough percentage of time in the selection of potential leaders to justify the spending of that much money on the selectees' training. He believes that industrial recreation programs are a more effective means of identifying and developing industrial leadership.

Earl Groves, former president of Groves Thread Company [his picture and statement preface chapter 1] feels that the identification of promising management material is one of the most important benefits of the recreation program at his company. In a taped interview with the author, Mr. Groves discussed his experience with the development of leadership quite extensively. "We are finding that most of our really promising young supervisors are those who came up through our recreation program," he said. And the inference should not be made that such persons are bound only for the lower order supervisory jobs. There is no limit to how high they can be promoted. At the Groves Thread Company, the executive vice-president is a product of the recreation program.

Mr. Groves looks for people willing to accept responsibility. He finds that "a high percentage of those who come in just want a job: 'Put me on a machine and let me put in my time, but don't burden me with responsibility.' "

Mr. Groves also believes in the recreation program as a means of identifying prospective supervisors, and he uses the program for that purpose, noting:

> Quite frequently we will check with our recreation director and ask him if he knows of someone who is suitable for a certain job. Very often he can come up with a person. We have found that a very high percentage of these persons work out very well. We know a whole lot more about them, and they have a much higher batting average than those selected otherwise."

The Groves Thread Company former president notes also that the recreation program sometimes develops persons who are not, at the outset, able to be placed in a position of responsibility. He sited one notable example:

> This particular young man came into our recreation program very young. His family background was such that his father was a dirt-track race driver, a pretty tough program.
>
> The young man was somewhat of a bully and constantly in fights; but, because of his desire to participate in athletics with us, he had to learn how to make himself acceptable to his peers.
>
> Eventually, he became an assistant to our recreation director and a volunteer coach and manager. On the strength of his performance, we hired him into our personnel department, and he did an excellent job there. We had an opening in the hosiery knitting department for a mechanic and he moved into that area, and again he did an outstanding job. He is now a productive member of the community.

The experience of the Groves Thread Company is not unique; it is just an example of what can be done in the way of developing leaders through recreation programs. Industrial recreation programs depend upon many volunteer group leaders and activities chairpersons. In this fashion they provide a laboratory for the recruitment and training of leaders.

Leadership qualities are often detected by recreation directors. It is also true that recreation activities sometimes provide the individual with the opportunity to discover in the self both the liking for leadership and the ability. Such a person is glad to take on other and more difficult responsibilities, being developed and tested simultaneously.

As one longtime recreation director noted, there isn't too much difference between selling an automobile tire than selling a person on participating in some activity. People who can effectively organize and manage recreation activities can also manage company functions. Recreation volunteers who are able to secure the cooperation of others have already demonstrated one of the most important qualities of a good supervisor. If one can get the most out of others in recreation activities, one can probably get the most out of those being supervised on the job.

Those who develop an interest in and show talent for leadership can also find in recreation clubs the opportunity to improve any areas of weakness. The employee who has other leadership abilities but has never been able to make a good oral presentation might, for instance, join the recreation organization's toastmasters club. The better recreation directors focus on young people of notable talent but little training. These persons can often be encouraged to attend classes at a nearby college and improve their mathematics or communication ability, or to take a course or two in management or in psychology. All that some of them need is the encouragement to get started.

Cultivating Leadership

Management particularly seeks integrity; and the recreation director has much opportunity to note among possible prospects for promotion, those who demonstrate personal integrity in recreational activities.

The identification, testing, and development of leaders is not some kind of covert operation on the part of the recreation director. Neither is it a case of subverting the recreational objectives to business purposes. First, it is necessary to identify and develop leaders in order to carry on a large recreation program. Second, the recreation participant benefits by being given the opportunity within the recreation program to—

  a. find out whether leadership is personally enjoyable,
  b. discover whether personally able to handle responsibility,
  c. gain experience in leadership,
  d. strengthen any areas of weakness,
  e. develop self-confidence,
  f. be promoted within the company, and
  g. have a better chance of success after promotion.

From the standpoint of the company, the recreation program has assisted in identifying, testing, and developing the quality which is probably most fundamental to the company's success—good leadership. Both the company and the individuals escape the trauma and the problems associated with the wrong selection of leaders. In the latter respect, Earl Groves said: "The wrong selection is very expensive, costing thousands of dollars; and, if they do not work out, you can't demote them. There is no choice but to let them go."

Fortunately potential leaders can be tested in the recreation program and, if found wanting, saved from mortification and dismissal. It is the best way for those who may succeed and the kindest way for those who would probably fail.

Community Service

In the present environment, there is much good to be derived from fun and games; but that is not the whole of life, nor is it the whole of industrial recreation. Among the important functions of the best of industrial recreation organizations, public service receives a significant amount of attention. This is fortunate, for, if this civilization is to survive, a great many individuals, as well as business organizations, need to develop and express a social conscience. It confers benefits to the individual giving the public service and is of great benefit to the company.

Areas of Public Service

There are many ways in which a company can render public service. One of the most obvious is through financial contributions, and that is important. In many instances the difference between the success or failure of a community enterprise is the financial support, or lack of it, by business and industry. However, as the president of Phillips Petroleum said, "The era is gone when an industry can consider its obligation to the community fulfilled by taxes and contributions to charity.[4]

A second and increasingly common way is for the company executives to spend a portion of their personal time on community affairs. In every community, there are businessmen who are outstanding public servants. Even if they were all so disposed, and a great many are not, the chief executives of American industry could give only a small portion of the volunteer time that is needed for the achievement of community objectives.

A third common way of contributing to the community is through encouraging the participation of company employees. As early as 1968, a Conference Board study shows that 87% of the respondents urged employees to volunteer their own time to community service.[5]

Employees and Community Service

As everyone knows, encouraging employees to participate and achieving their actual participation are two different matters. Even today, most employees—even those of such presumably public-minded institutions as schools and universities—do not participate extensively, if at all.

When there is a well-led industrial recreation association, there is a different story. Employee cooperation on community projects which the company wishes to support can be gained wholeheartedly through the recreation association. The recreation association is ever so much more effective as a motivation device than is a letter from the company president, messages in the company newspapers, or announcements on bulletin boards. It is more effective than the intervention of the personnel department or the thinly veiled "suggestions" of official channels. If, as a result of such channeled suggestions, the employee does "cooperate," it is often halfheartedly and with resentment. Many recreation clubs voluntarily, and as a part of their objectives, participate in community projects.

Recreation Groups in Community Service

Industrial recreation groups are superior in many aspects of community service because they are already organized. They have a strong sense of group identity. They are accustomed to working together. They know where to turn among other employees for special services or expertise. And they are desirous of creating a good image for their organization and the company.

An unusually good illustration of the power and effectiveness of an organized recreation association participating in community service is offered by the Goodyear Tire and Rubber Company employee recreation organization at Akron, Ohio. It has more than forty-five different activity clubs and many of them serve the community as part of their objectives. At the nucleus of community efforts are the one thousand volunteers, many of them club officers. The various recreation subclubs have contributed time, talent, and money to a great many community programs including:

> Assistance to handicapped children
> Children's Home vocational scholarships
> Juvenile Court chapel needs
> Coronary monitoring units
> Shock therapy carts
> Cerebral palsy units
> University fellowships
> Blood donor programs.[6]

The money contributions have been matched by many hours of volunteer service. This kind of service is meaningful recreation for many persons and is the way they choose to use much of their leisure time.

In mentioning Goodyear and public service, we cannot pass over the Boy Scouts. In 1914 Goodyear sponsored a scout troup for sons of employees at Akron. They met in a plant conference room. Today the Goodyear Tire and Rubber Company is the largest industrial sponsor of scouting. In 1973, it sponsored 67 units with 2,200 scouts. Needless to say, many of the volunteer workers and leaders of scout troops and activities are members of the recreation organization.[7]

The Goodyear Scouting Lodge in Akron, Ohio. Valued at approximately $200,000, it demonstrates Goodyear's commitment to Scouting.

Another important way in which a company can make important community contributions is through the sharing of recreation resources with the community. This is often most feasible and helpful in communities of relatively small size, such as Bartlesville, Oklahoma (Phillips Petroleum Company) and Bensenville, Illinois (Flick-Reedy Company). Presidents W.F. Martin of Phillips Petroleum and Frank Flick of Flick-Reedy are strong supporters of industrial recreation and of their communities. Those interests mesh beautifully.

In each instance to be mentioned, much industrial recreation association volunteer help goes along with the facilities. The volunteers and the facilities constitute outstanding community service. Among other functions, of which there are many, Phillips Oil Company uses their fine pool facilities to host high school swim meets and also national AAU swim meets.

At the Flick-Reedy Company, approximately 1,400 children participate in a Red Cross Learn to Swim Program every summer; and, from September through June, swimming instruction and "a lot of tender, loving care" is given to hundreds of the mentally or physically retarded. In an address presented at the 30th National Conference of NIRA, Frank Flick, the company president, revealed: "With the assistance of volunteers, we have helped a lot of these handicapped kids develop their neuromuscular coordination—in our pool and on Exer-cor [an exercise device invented by Mr. Flick]. And this is only a partial description of how our recreation facilities and personnel serve the community."[8]

The examples given are far from unique. There are a great many instances where industrial recreation programs perform similar community service programs. Through such means, employees achieve a high degree of self-fulfillment while contributing to the corporate citizenship role and enhancing the image of the company.

Social Service

There is one special way in which industrial recreation contributes to the public good: that is in the things it does to help stabilize and maintain the family as an institution.

Scholars have noted often that primary institutions are losing their vigor. Membership in organized churches has declined in recent years, and many of those who are on the rolls officially actually do not participate. We are predominantly a secular society.

The public schools, once conceived as the mainstay of society, have come in for a great amount of criticism, both in terms of their effectiveness as basic teaching tools and as transmitters of our social heritage. Many school teachers themselves disavow their former roles as molders of character. The effective socialization of children is sort of a no-man's-land, inasmuch as many parents leave that to the schools while the schools insist it should be done at home. And churches, for the largest part, are of only minimal influence on our growing children. The social problem is worsened by the heavy incidence of divorce and broken homes. Broken homes contribute to juvenile delinquency and the drug culture.

Acknowledging differences of opinion in some quarters, most of us feel that the preservation and strengthening of the family is of the utmost importance to society.

The family in America is changing both in its form and its basic functions. Recreation as an important family function seemed to be declining into insignificance a few years ago, but now it is beginning to reassert itself.

Industrial Recreation and the Family

Industrial recreation brings the family together in play, and that is important today. This is emphasized by the sociologist Judson Landis, who explains that, as other functions of the family have become less important, play has become more important. It is the means of encouraging families to do some things together, and it provides parents with the opportunity for demonstrations of interest and affection. Play, he says, "is approaching equivalent importance with the provision of emotional support and socialization. Indeed, play is becoming a chief mode of the emotional-support and socialization functions of the family in our society."[9]

Not a sociologist, Henry Ford II echoes some of the same ideas about play and the family in his statement from *Top Management Speaks* which prefaces chapter 4.

> Play is an important and necessary part of good living. It is humanizing. It reaches across the barriers of age, custom, and discipline, so that father and son get to know each other as people. It adds warmth and pleasure to family life.

These are some of the reasons why industrial recreation leaders and the sponsoring companies place such emphasis upon family participation. We illustrate this common concern with four brief quotations, all parts of longer statements in *Top Management Speaks* and used in chapter prefaces:

> Raymond Herzog, 3M Company, this chapter: "Our emphasis . . . continues to be on family participation."
>
> Robert W. Galvin, Motorola, chapter 8: "Employee recreation is of great value to families."
>
> Frank Flick, Flick-Reedy, chapter 11: "Now it [industrial recreation] is being broadened to embrace the family, with more emphasis on family programming."
>
> Robert M. Hoffer, Wisconsin Gas Company, chapter 4: "In order to be most successful, a recreation program must make provision to include the employee's entire family."

Family Service

Recreation activities are the chief source of industrial recreation influence on the family; but services to families are a very important consideration in many programs. Such family services include: day-care centers, summer camps for employee's children, money-saving group tickets to recreational programs and amusement parks, and employee-operated stores, where family shopping is made more convenient.

Company Public Relations

The best company image and the best community relations are developed by community service. Each community is different and requires different understanding; but the local prestige of a company is everywhere going to be correlated with the extent of company participation in community events and in community service. Management can do a great deal; but its efforts can be magnified many times over by the activities of the members of the recreation association. Industrial recreation cultural activities are also beneficial to the company's public image. Consider, for instance, the public relations value of a sixty-piece band which plays to fifty thousand people a year. Many industrial recreation organizations have bands, orchestras, or theater groups which give public performances in their own communities and, sometimes, at quite some distance from

Always remembering the children at Christmas time, recreation associations almost all have parties for the children. Here is Santa at the Battelle Laboratory party and at the Pratt and Whitney Aircraft Club party.

Family outing (child tossing ball).

Family outing (tug of war).

Family outing (sack races).

Children's circus.

Pratt and Whitney Aircraft Club demonstrates interest in the family.

home. They are often composed of quite talented persons, who develop a good deal of group pride, and put tremendous effort into producing shows of which they, the recreation organization, and the company can be proud.

Company Salespersons

In the long run, the company's most effective salespersons are its own employees *if* the employees are pleased with their company and personally identify with it—as most likely will be the case if there is a strong and well-supported recreation program. A company with three thousand pleased and happy employees has three thousand voices of good will. That number may be doubled or tripled by the employee's family—especially that will be likely if there is a family-oriented recreation program. This is not to mention the employees' friends who may be so impressed with the good things they hear about a company that they also become agents of good will in the community.

Employee Morale

Employee morale is a large topic about which whole books are written. It is difficult to handle in a short space because there is so much to be said and so many factors to be considered. Its various aspects are both separable and inseparable. Though one can recognize specific aspects which can be discussed, no aspect can be discussed out of context with other aspects. For instance, morale is closely tied in with communication, discussed earlier.

We approach morale basically in terms of the well-being of the individual, but we recognize that the well-being of the individual has important, though largely intangible, effects upon the company in terms of public relations, productivity, and so forth.

Positive Morale

The feelings of people at work are a mysterious power. Money can buy human energy, but it cannot buy enthusiasm! Nor good will! Nor team spirit! Enthusiasm, good will, and team spirit combined seem to act like some kind of magic which lubricates the machinery, causes supplies to flow, makes the motors hum, regulates the gears, meshes the cogs, and converts the whole into quality products at the end of the assembly line.

Many industries are largely automated. They will be more so in the future. But no amount of automation, no amount of sophisticated mechanical precision, and no amount of computerized control—nothing will reduce our dependence upon human enthusiasm, good will, and team spirit.

On the contrary, the greater the sophistication of automation, the more dependent we become upon human attributes for meticulous input, maintenance, and repair. Do not even think of deliberate malicious acts, just shudder at the monumental foul-up that can result from one moment of carelessness!

Outside of manufacturing and industry, the examples will be different, but the effect of morale will be the same. Consumers and work orders are a nuisance to those of low morale, but a challenge—more, the opportunity for earning a living, for personal achievement, and promotion for those of high morale. The human input in all kinds of businesses, industries, and governmental units is just as significant. Computers and robots may someday *run* the world but they can never *direct* it.

Negative Morale

Sometimes the person who cannot recognize the contribution that high morale makes to productivity does understand the negative effect of low morale on production.

Where there is notably low morale, the workers are preoccupied with pay, vacations, and coffee breaks. They are interested not in what they can produce, but in what they can get.

Inevitably our attitudes affect our behavior, sometimes unconsciously. The unhappy employee is more likely not to practice good preventive maintenance—not to lubricate thoroughly, not to protect from rust, not to guard against damage, not to be careful about the quality of the product or the satisfaction of the customer.

Low morale makes employees careless and not safety conscious. It may be that, unconsciously, they court injury as an excuse to stay off the job and draw sick leave, thus "punishing" the employer.

Low morale makes interpersonal relations more difficult. The resentful employee may ignore or disobey orders, and, more than likely, will quarrel with superiors, peers, and inferiors. The resentful employee often becomes a thoroughly disagreeable person, whose mood can adversely affect many other workers. At the greatest extreme, the disgruntled worker will deliberately engage in sabotage of machinery or product, commit arson, or physically attack a work associate. Low morale of any degree is a drag upon production!

Morale and Industrial Recreation

Unquestionably there is a correlation between morale and productivity. Is there also a correlation between well-planned and intelligently directed industrial recreation programs and morale? We think so! We sincerely think so for these four reasons:

1. Industrial recreation programs logically should affect morale.
2. Industrial recreation meets many of the recognized human needs.
3. Industrial recreation is appreciated by employees.
4. Industrial recreation is believed by experienced employers to have a good effect on morale.

Human Needs

Psychologist A.H. Maslow is associated with the theory that human motivation is on a continuum and can be arranged on a hierarchy of needs, from lower and stronger to weaker and higher. His is an optimistic approach to human motivation, is reasonably consistent with our own experience, and has met with a good deal of acceptance. Industrial recreation can easily be related to other psychological theories, but we are inclined to that of Maslow.

In brief and, we hope not oversimplification, Maslow's first five steps in the hierarchy of human needs are:

1. physical needs—food, rest, sleep, sex
2. safety needs—security, freedom from danger, desire for the familiar
3. belongingness and love needs—desire for group identification, wish for acceptance
4. esteem needs—need for achievement, status, recognition
5. need for self-actualization—self-fulfillment, the need to become all that one is capable of being.

The idea is not that industrial recreation is essential to any of those needs or completely satisfies any of them, but that it makes some contribution to all of them, some more than others.

1. *Physical needs.* Among these is the need for activity, which is met by sports and other activities of recreation. The need for relief from tensions, which recreation provides, would also fit into this category.

2. *Safety needs.* Employees may find in the industrial recreation organization a sense of security. By finding recreation with fellow employees in familiar settings, one meets the security that comes from familiar surroundings and familiar faces. Moreover, there is a real sense of being protected by the larger group.

3. *Belongingness and love needs.* Employees find the opportunity to identify with the recreation organization as a whole as well as with separate clubs. It is important, especially in large companies, that one feels accepted by the recreation group.

4. *Esteem needs.* Industrial recreation presents many opportunities for leadership and for accomplishments which can bring recognition from others as well as self-respect.

5. *Self-actualization.* The individual finds opportunity to improve to the utmost of capability through the development of new skills and through cultural growth. Through such community service projects as already described, the individual has much opportunity to find self-fulfillment through doing for others.

Appreciation of Industrial Recreation

Most employees do appreciate the opportunity to participate in industrial recreation, whether or not they take advantage of it. There is the Hawthorne effect: believing that something is being done for them, they react with good feelings about the company and everyone concerned.

There have been several research projects concerning employee acceptance of industrial recreation. Unfailingly, research has indicated positive attitudes toward industrial recreation.

A master's thesis research study of "Industrial Employee Attitudes toward the Values of Industrial Recreation," completed in 1971, concluded that "both participants and nonparticipants hold a positive attitude toward industrial recreation values . . . although the participants hold more intense positive values." An interesting and unpredicted conclusion was that "union members disclosed a greater intensity of attitudes than did nonmembers."[10]

A research project at the Johnson Wax Company in Racine, Wisconsin, was conducted by Dr. Reginald Carter. His findings served to reinforce the conclusions of earlier researchers in respect to employee acceptance of industrial recreation. Most of the employees surveyed reported "high satisfaction" with the recreation program, and a total of 92% reported at least "satisfaction."[11]

Experienced Employers

In turning to the opinions of experienced management, we refer to those represented in *Top Management Speaks.* It would be very simple to find many others. Those in *Top Management Speaks* are highly placed executives who have had long and close familiarity with industrial recreation in their own companies. It is logical in researching customer satisfaction to "ask the man who owns one."

Obviously all of the eleven do believe in industrial recreation, but it is notable that ten of them, in rather brief statements, make some reference to morale or the factors that contribute to morale. If you read them all carefully, you will be surprised to find how many references there are to the human needs that Maslow identifies. The two for this chapter both indicate human needs met by their recreation programs:

Raymond Herzog, 3M Company:

> The personal friendships formed with fellow employees from all areas of the company contribute to a feeling of belonging and help promote the small company atmosphere that could diminish as a company grows.

Robert W. Galvin, Motorola incorporated:

> The basic objective of industrial recreation is to recognize man's needs as a social entity. . . . Employee recreation has given opportunity for personal expression, individuality and recognition to the men and women in industry.

Indeed, it is logical that industrial recreation should improve morale—it does meet many of the recognized human needs, it is appreciated by employees, and, employers who have had experience with it believe that it improves morale.

The values of physical fitness, the next topic, are also easily established.

Physical Fitness

Physical fitness, which will receive more attention in chapter 7, is briefly discussed here because of its relationship to morale and productivity. The President's Council on Physical Fitness and Sports defines physical fitness in these words:

> [Physical fitness is] the ability to carry out daily tasks with vigor and alertness, without undue fatigue and with ample energy to enjoy leisure time pursuits and to meet unforeseen emergencies. Thus, physical fitness is the ability to last, to bear up, to withstand stress, and to persevere under difficult circumstances when an unfit person would quit. . . . Physical fitness is more than "not being sick" or merely "being well." It is different from immunity to disease. It is a positive quality, extending on a scale from death to abundant life.[12]

Benefits of Programs

Physical fitness programs, initiated by the U.S. Marine Corps, were subsequently adopted by business and industry because of the numbers of deaths from heart attacks of young executives in positions of stress. In some companies, the physical fitness program is only for executives and is rather preventive health measures than recreation. Yet even in those companies it is commonly under the direction of the recreation department. Increasingly, however, as the rank and file workers saw the benefits and requested the opportunity to engage in such activities, physical fitness programs were extended to all employees. Today physical fitness programs are important parts of many employee recreation associations.

There is a wealth of research in the United States and abroad which corroborates the value of physical fitness programs. There is complete agreement that a feeling of well-being, an established product of physical fitness program activities, improves morale. We note here only a few of such research projects.

In 1968, the National Aeronautics and Space Administration (NASA), in cooperation with the U.S. Public Health and Heart Disease Control Program, conducted the first major study of the effects of physical fitness programs in American business and industry. The extensive report concluded that those who persevered in the program improved their work performance and held more positive attitudes about their jobs.

Similar results were found in a study in 1970 by Fred Heinzleman which was supported by the Heart Disease and Stroke Control Program and reported in the October 1970 issue of *Public Health Reports.*

Dr. George Otott, Director of the Southern California Universal Recreational Fitness Research Staff, reported that a fitness center not only benefits it users "but it can also be an excellent way for the employer to increase the morale of his people while improving their health, fitness, and productivity."[13]

There seems almost no question but that physical fitness programs do improve the health of employees as well as their morale. Where physical fitness is an objective of industrial recreation programs, good results will follow.

Recruitment and Retention

That industrial recreation programs really improve recruitment is probably almost impossible to prove by hard data. Yet research studies strongly suggest that conclusion, and there is other evidence that supports the belief.

There are many statements by management indicating the belief that recreation programs are of assistance in recruiting. Raymond Herzog, in his statement which prefaces chapter 3, asserts: "We feel that the 3M program exerts a strong, positive influence on the quality of employees we attract and retain." This benefit is also mentioned by others of the top executives.

Among the best evidence that we have found is that which is furnished in a tape-recorded interview with Earl Groves, former president of the Groves Thread Company of Gastonia, North Carolina. His words are most persuasive.

> We have an industrialized county with over a hundred textile plants. As we became more and more industrialized, we experienced a rather intense labor shortage that started in 1968 and persisted right up until the recession.
>
> In Gaston County we reached a low of .3 of 1% of unemployed. We had a good many unfilled jobs in our county because when you get much below 3% of unemployment, you have virtually no employable people available. A lot of people are classified as employable that really are unemployable, at least by plants like ours.
>
> During this period we came to realize that while we were suffering the effects of the labor shortage, we were not being hit nearly as hard as some of the other companies. Other companies came around inquiring—wanting to know why we were attracting more competent employees, wanting to know the values of our recreation program.

The experiences related by Mr. Groves are not unlike those we have learned from other sources, though not so directly. The Goodyear company at Akron has reported similar results, particularly in respect to its support of scouting, which is one aspect of its recreation program. As it is related in one article, "A large number of Scouts join the Goodyear Company after completing their education. Board Chairman [1970] Russel De Young is a veteran of scouting and many other employees are former scouts."[14]

To a considerable extent, recruitment and retention go hand in hand and what promotes the one promotes the other. It is also probable that absenteeism may be correlated with low morale and hence is also correlated with the lack of industrial recreation programs.

Many persons are skeptical that industrial recreation can reduce absenteeism, yet there is evidence that it does. There are two similar reports concerning the Diamond Alkali Company. That company introduced a recreation program particularly to combat absenteeism at its Deer Park Caustic Chlorine Plant near Houston, Texas. Employee help was donated to build the facilities and group spirit increased as work progressed. The company was persuaded that its recreation program did reduce absenteeism markedly. According to the reports, labor turnover also decreased.[15]

A study by the National Industrial Recreation Association, reported in the book *Is Anybody Happy?* in 1962, indicated that the employees of companies with recreation programs had a lower incidence of absenteeism and a higher rate of job efficiency and morale.[16]

A master's thesis study at Pennsylvania State University reported that at Admiral Corporation, a highly mechanized plant that used many unskilled laborers, absenteeism became a problem and production suffered. After the management introduced a recreation program for the employees, absenteeism decreased and production climbed to a new high.[17]

Actually, it is difficult not to believe that there are days when some employees go to work because of some activity sponsored by the recreation program, or because of a responsibility that fell to them because of their role in the program. It is just as easy to accept the idea that employees sometimes consciously or unconsciously make a decision to stay home from work because of morale problems or psychosomatic illness. Positive and negative approaches suggest that there almost certainly is a correlation between absenteeism and the presence or absence of a good recreation program.

Retirement

Naturally, it is the company pension plans and Social Security that have the most to do with the financial welfare of retirees. However, it is as true of retirees as of any other age group that man does not live by bread alone. All of us have witnessed the terrible futility of retirement for many workers, especially men. Too many are like the actual case of a man who said, "All my life I looked forward to retirement because I did not have the time to do many of the things I really wanted to do. Now that I am retired, there is nothing I want to do."

Preparation for Retirement

As mentioned earlier, a great deal is done by industrial recreation programs to prepare persons for retirement. That is a deliberate objective of many programs. The important thing is to develop interests and skills during the work years that may sustain one's interest after retirement. With recreation programs which have a wide range of hobby clubs, that is easy to do.

Keeping Contact

Most industrial recreation groups continue the membership of retired employees and make special provision for them in the group activities. Retirees and families are also made welcome at recreation facilities. It is very important that the retirees not be shut off from their friends at work, as though a door were to close. That is what happens where there is not recreation association.

The number of retirees is growing and will grow for the foreseeable future. All industrial recreation groups need to make preparation for retirement an important objective. Those who do confer a benefit on the company as well as the retirees.

Conclusion

With the benefits of retirement-preparation, we conclude the first part of this book. As well as we could within the confines of these pages we have tried to make clear why industrial recreation exists in terms of its characteristics, the problems and needs of capitalism and society, and the benefits conferred to industry and the public by industrial recreation.

Part II will be concerned with how industrial programs are organized and operated.

Notes

1. "Studies in Personnel Policy, No. 200. A Research Report from the Conference Board," 1966, p. iii.
2. Ibid., p. 11.
3. Reginald Carter, Ph.D., and Robert Wanzel, Ph.D. "Measuring Recreation's Effect on Productivity." *Recreation Management,* August 1975, p. 43. *Recreation Management* is the professional journal of the National Industrial Recreation Association. It will be quoted innumerable times in this book. Hereafter, in footnotes, it will be referred to as *R.M.*

4. W.F. Martin. "The Case for Company-Community Relations." *R.M.,* March 1972, p. 12.

5. "The Role of Business in Public Affairs." Studies in Public Affairs, No. 2, 1968, p. 25.

6. Esther Winchell. "Something for Everyone." *R.M.,* October 1973, pp. 12–13, 22. Also, Charles Bloedorn to Theodore Wilson, 17 June 1976.

7. Ibid.

8. Frank Flick. *The Untapped Potential: Industrial Recreation.,* n.d., published by NIRA, Chicago, Illinois.

9. *Sociology: Concepts and Characteristics.* (Belmont, Cal.: Wadsworth Publishing Company, 1974), p. 357.

10. Jerry Rogers Garden (Master's Thesis, Pennsylvania State University, 1971), pp. 69–71.

11. Reginald Carter and Robert Wanzel. "Measuring Recreation's Effect on Productivity." *R.M.,* August 1974, p. 42.

12. "Physical Fitness Defined." *R.M.,* April 1973, p. 32. Reprinted by permission of the President's Council on Physical Fitness.

13. "Recreational Fitness Centers: The Time Is Now!" *R.M.,* April 1972, p. 33.

14. Winchell. "Something for Everyone." *R.M.,* October 1973, p. 14.

15. Bennet Berger. "The Sociology of Leisure: Some Suggestions." *Industrial Relations.* February 1962, pp. 31–45. "Playing at the Plant." *Business Week,* 20 February 1954, pp. 74–80.

16. Norman M. Lobenz. (Garden City, N.J.: Doubleday, 1962.)

17. Jerry Rogers Gardner. "Industrial Employee Attitudes toward Industrial Recreation," 1971, pp. 14–15.

PART II

4
Organization

TOP MANAGEMENT SPEAKS

A popular feature in many American homes is the place where the family gathers to relax and just have fun together. It may be a room set aside for that purpose, complete with television and ping pong table, or perhaps a corner of the living room with a shelf to hold books and games.

Here an important family ritual is performed as children and parents and friends put aside the problems of the day to share a common and pleasant experience. Play is an important and necessary part of good living. It is humanizing. It reaches across the barriers of age, custom and discipline, so that father and son get to know each other as people. It adds warmth and pleasure to family life.

By the same token, *there is an important and valid place for recreation activities in industry. This is by no means a paternalistic viewpoint.* Providing suitable recreational outlets for employees helps to give employees a sense of common identification and to broaden their pleasure and satisfaction in their work. *In a direct, practical and forthright way, it expresses the company's concern for the human well-being of its employees and its gratitude for their essential contribution.*

The benefits of this program to employee and company alike are intangible. They cannot readily be measured in dollars and cents. But I am sure they are substantial indeed.

Henry Ford II
Chairman of the Board
Ford Motor Company

Henry Ford II

TOP MANAGEMENT SPEAKS

Wisconsin Gas Company has long recognized that a well-rounded recreation program is a key factor in developing and maintaining the morale of its employees. We feel that this program is one of the best investments we can make in providing a means for our employees to get to know one another outside of the work environment.

Recreation programs are designed by and for the employees. We attempt to tailor our programs to fit the expressed desires of those who will be participating in them. Whenever there is a significant expression of interest on the part of our employees in a particular activity, we try to be responsive and place this activity in our program.

There is no exclusivity in any of our recreation activities. The members who participate are from all areas of the Company: shop, office, and management. It has been our experience that the mutual interests and understanding fostered by participation in recreation programs carries over into the business life of the Company, and provides a cohesiveness that is highly desirable.

In order to be most successful, a recreation activities program must make provisions to include the employee's entire family. Many of our recreational programs are structured toward encouraging participation by spouses and children of employees. Our family bowling functions, group outings, and annual children's events have been particularly successful in this area.

Robert M. Hoffer

We believe that employee recreation activities provide a vital part of our human relations program. Although the dividends these activities pay cannot be measured in dollars and cents, they are invaluable in terms of morale, employee satisfaction, and the ultimate success of the Company.

Robert M. Hoffer
President
Wisconsin Gas Company

With this chapter we begin the "nuts and bolts" of the book but it is not at all like a mechanic's manual. It will not tell you how to assemble, disassemble, or repair anything. It is not intended to be a handbook for ready reference about how to do this or that.

Essentials but Not Specifics

If you will pick up *The Joys of Cooking* and turn to page 75, you can find explicit instructions for the preparation of grilled lobster tails. If you prefer Scandinavian Krumkakes, the exact list of ingredients, the proper amounts of each, and directions for mixing may be found on page 375. But nowhere in this book will there be listed the specific ingredients, much less the exact amounts and directions for blending them.

Recipes and explicit instructions are omitted for two reasons. The first is that we know of no "sure-fire" recipes. Industrial recreation is a social institution. It is made up of individuals, each one unique. Moreover each recreation organization exists in a unique social environment or climate in a unique moment of history.

The second reason is pragmatic. One has to work with what there is. If any of us were to know the ideal organization for a specific program in its social setting, it is quite unlikely that we could establish it in its perfection. It is probable that either Management; the vocal, self-appointed leaders of the employees, or the general groundswell of opinion among the employees would reject our ideal solution.

Certainly no one should expect to find among the different organizational forms to be discussed one which will be ideal for your recreation organization. The organizational form should flow out of the interrelationship of the unique circumstances of each company—the basic philosophy, the long term goals, the nature of the present operation, the employee makeup, and a host of other factors. If one organization tries to adopt an entire pattern which appears to work well elsewhere, the organizational form may operate like the fabled Procrustrean Bed.*

The last thing we want to suggest is standardization or operation by the numbers. All that we aim to do is make clear advantages and disadvantages of different organizational patterns in different circumstances.

At least in the past, very few, if any, industrial recreation programs have sprung into being full blown. A few of them have begun as the result of management's decision to provide more benefits to the workers. Most have developed through the growth of employee interest in special activities.

Origins of Industrial Recreation

Some employee participation usually comes about even before the dust of construction of a new plant has settled. A few employees eat lunch together and begin throwing a ball around. Before long a team has been formed. Someone suggests that the team have a name or identification, and the most natural development is identification with the company and the company name.

After that someone usually approaches Management about the use of the company name or the use of company bulletin boards or news publications for publicity. Often they want to use the

---

*Procrustes was a celebrated legendary highway man of Greek mythology. Procrustes tied his victims to his iron bed and, as the case required, either cut off their legs or stretched them until there was a perfect fit.

company name on their uniforms and it occurs to someone that the company might be willing to furnish the uniforms and underwrite some of the team expenses.

Almost always someone in management, probably in the personnel office, feels the desirability of company involvement, if for nothing else than for the sake of being kept informed.

The actual history of the large and flourishing Lockheed Employees Recreation Club is a typical example of such a beginning. In April of 1935, a small group of employees discussed the advantages of coordinating the various small group activities within the company. A committee set to work to draw up a constitution and a small delegation was sent to ascertain the attitude of Management. The project was encouraged by one Lockheed vice-president who donated twenty-five dollars out of his own pocket for initial expenses.

A second meeting was held in May, 1935, and the constitution and bylaws were approved. On July 22, 1935, the first regular meeting was held. This meeting was attended by representatives from each work department and officers were elected. By year's end much progress had been made, and the Lockheed Employees Recreation Club (L.E.R.C.) had raised enough money through the sale of candy to have the first Christmas party.

During the next three years, the L.E.R.C. grew rapidly both in number of members and in the diversity of its activities. On April 18, 1939, it became a nonprofit corporation.

The Need for Rules

Whatever the beginning of a recreation organization, the need for rules of operation become apparent quickly. There must be some means of determining authority in order that decisions may be made. The organizations may take myriads of forms, and the possibilities are so great that no two are exactly alike. Nor do they need to be. A cardinal principle of industrial recreation is that "except for the centralization of responsibility and the inclusions of employee participation, industrial recreation has no universal form, habitant, situation, status, or activity."[1]

The Basic Rules

Different as organizations may be, they are alike in requiring some basic rules, variously known as the constitution, bylaws, or both. Because such basic set of rules are common to all kinds of organizations and information about them is available from many sources, the discussion here will aim at brevity.

There appears to be confusion about the terms "constitution" and "by-laws." There need not be for they mean the same things, that is the basic rules of the organization. Many organizations have both a constitution and a set of bylaws. Unless much care is exercised, there may be some contradiction between the two, and, if there is none in the beginning there almost certainly will be with passage of time, as one or the other is revised without a realization of the effect upon the other.

George Webster, an attorney and one of the leading authorities on the subject believes that where bylaws and a constitution already exist, they should be merged into one. He favors the term "bylaws" because, he says, "its legal meaning possesses greater definition and understanding."[2]

For simplicity, we shall hereafter refer to the basic and guiding rules as bylaws, though we think the choice of the one or the other is less important than that there should not be both. It is essential that the bylaws be characterized by simplicity and specificity. This will help to avoid confusion and all manner of internal dissension.

Bylaws: Form and Suggestions

The following outline, with some suggestions, is not intended as a model which will be suitable for all organizations. It should furnish food for thought, which is all that is intended. For every suggestion there will be some exceptions.

I. Name of Organization

Even in this simple matter, some thought should be given. If the association intends nothing more than sports and social clubs, the term "recreation" is suitable. If services, such as group purchases of tickets or merchandise discounts are intended, the word recreation is not sufficiently descriptive. Moreover, we observe that many groups which began merely as sports programs later branched out into services of various kinds. This expansion of activities is one mark of maturation. We would suggest, therefore, that the name be more descriptive. "Activities" is suitable even though the only activity is a softball team, and "activities" remain appropriate in the event the association eventually broadens its objectives to include other kinds of activities and services.

It would help to change the image of the entire field if newly formed groups should choose "activity" rather than "recreation" as the descriptor in their names.

II. Objectives

Clarity is essential and specificity is highly desirable. Objectives so general that they will fit virtually all similar organizations are a waste of time. The organizers should list the specific intended objectives. Nevertheless it is wise to note that the objectives listed are neither all-inclusive nor mandatory.

III. Membership

A. Full Membership

Full membership should be open to all employees of the company or companies concerned.

There will be instances when company executives, whether because of company policies or their own biases, do not participate. There are many companies and many executives who believe that executive socialization with the rank and file employees weakens supervision effectiveness on the job. In the article, "Should You Socialize with the Troops?" B.B. Carrol concludes "when you balance the pros and cons, the argument is much stronger against it."[3]

It is true that the weak executive should not socialize for the weakness will be discovered. The really able executive need have no fear of being exposed to all situations.

There has been a slow but persistent growth of social and economic democracy in America since the Jacksonian Period, now more than a century past. Yet there are those who fail to note the tide until it engulfs them. There is no room in American industry today for Officers' Clubs and Enlisted Men's Clubs. Fortunately both for industry and employees clubs, a growing number of highly placed executives are of the persuasion expressed by Robert M. Hoffer in *Top Management Speaks.*

> There is no exclusivity in any of our recreation activities. The members who participate are from all areas of the Company: shop, office, and management. It has been our experience that the

mutual interests and understanding fostered by participation in recreation programs carries over into the business life of the Company, and provides a cohesiveness that is highly desirable.

B. Honorary Membership

Most employee asociations choose to continue retired employees as honorary members. If there are special retiree clubs or activities, retirees might therein be as full members with the privilege of voting and becoming elected officers.

III. Members' Responsibilities

It will highlight the privilege of membership and increase the sense of its worth if members have two responsibilities: to pay dues, however small, and to sign a membership card.

General dues may be small in order not to discourage membership. They may be token rather than fund-raising. As has been often observed, people place more value upon things which cost them something. If dues are needed for revenue, it may be wiser to place charges for participation in activities.

IV. Membership Privileges

Privileges include the right to participate in association activities and services of all kinds. Additionally they include the right to vote for officers in the association and to be considered for officers.

V. Meetings of the Membership

A. Regular Meetings: Regular meetings of the membership should be sufficiently infrequent that there will most likely be matters of significance upon which to take action. Meetings which consist of only announcements and reports discourage attendance.

B. Procedures: Accepted parliamentary procedures are a must. Robert's Rules of Order have been tested by time and are recommended.

The bylaws should state what percent of the membership must be present to take official action. Circumstances vary widely, but there is a greater danger of setting the necessary percent too high than too low.

C. Special Meetings: The bylaws should state the authority for calling of special meetings. The bylaws should also state how much notice must be given for special meetings and when the agenda must be published. No official action should be taken upon matters not listed in the agenda.

VI. Governance

A. Functions of Governing Board

Common responsibilities of the governing board include: .

    a. To develop and review policies
    b. To make *critical* decisions or to recommend decisions to the whole membership
    c. To coordinate the activities and efforts of the different divisions or clubs
    d. To serve as liaison between the organization and Management
    e. To inform and counsel with the administration
    f. To review administrative actions
    g. To receive and act upon financial reports
    h. To review the bylaws and objectives, and perhaps,
    i. To appoint certain officers

Notably the board should not attempt to administrate. That, it cannot successfully do. It should not become involved in the day-to-day administrative actions. It can decide the destination but it cannot drive the car.

Governance and administration are different. The responsibilities of the administrator are discussed in chapter 8. It is highly important that the governing board and the administration be cooperative and supportive of each other.

B. Membership of Governing Board

The Board should be relatively small in numbers, certainly not much if any more than a dozen. The larger the board the more inefficient it will be in terms of the time necessary to reach decisions. If for the sake of good representation, the governing board needs to be large, it should elect an executive board from its own members and have infrequent meetings.

In almost all cases, the governing board should include at least one management representative, whether ex officio or with rights. That is essential for good communication and cooperation.

The length of governing board terms should be long enough that the members have time to profit from the experience they gain. It usually works well for board members to serve two-year terms and to provide that the terms of members be staggered. That way half of the board will have had at least one year's experience. The staggering of terms also leads to greater continuity.

VII. Officers of the Association

The qualifications of the officers is more important than their personal popularity. Experience is also important and many organizations provide that the vice-president, or the first vice-president if there be more than one, succeed automatically to the presidency. A desirable qualification of all elected officers is that they should have had previous experience on the board or in some other important capacity.

In some cases, as the treasurer of a large organization, expertise is so important that the position ought to be appointive rather than elective. The same may be true of the secretary.

For every officer there ought to be a clear job description. Good job descriptions make it more likely that the work will be done and unnecessary rivalry will be avoided.

VIII. Committees

    a. Functions and powers

    b. Membership and method of selection

    c. Length of members terms. Again staggering may be wise.

IX. Fiscal Details

    a. Responsibility for disbursing funds

    b. Accountability for funds

    c. Fiscal auditing: Annual auditing by a company or outside auditor is a wise provision where large sums of money are involved.

X. Amendments to the Bylaws

The bylaws should be neither carved in granite nor scratched in sand. Amendments should not be so difficult as to approach the impossible nor so easy that they may be made upon whim. The provisions for amendments are among the most important of the entire set of bylaws. In keeping with their importance we would make a few recommendations.

1. Amendments should be the perogative of the whole membership only.
2. Proposed amendments should be discussed in at least one general meeting or in a meeting of the governing board open to the whole membership before a vote is taken.
3. Proposed amendments should be published to the whole membership.
4. Voting should be by secret ballot.
5. Every member should have the opportunity to vote, which precludes voting during meetings.

6. After publication and discussion, adoption should be by a predetermined portion, perhaps two thirds, *of those voting.*

It is important to stipulate "of those voting" so that disinterested members not make amendment impossible.

The Importance of Bylaws

Unless adhered to, bylaws are worthless. If, in practice bylaws are found to be unworkable or unwise, they should be revised. Provision for a regular review of objectives and bylaws is wise. Sometimes such a review may result in a revision of the bylaws. At another time the result may be revision of operations to bring them into line with the bylaws.

In the vast majority of cases it is better carefully to revise the bylaws than to throw them out and start all over. The brand new bylaws will probably cure the problems which brought them about, but they will almost surely create new problems of their own.

Relations of the Association to the Company

The essentials of bylaws are easier to define than is the content. In regard to the content there are many things to consider. An important consideration is the relation of the employee activity association to the parent company. Theoretically this relationship may range from absolute company control to complete employee independence.

Actually the degree of control is on a continuum, and we cannot talk so much of company control or employee freedom except as a matter of degree. We can discuss the disadvantages of both extremes. On the suspicion that more employee programs lie nearer to company domination, we begin with that extreme.

Disadvantages of Company Domination

1. The greatest disadvantages of company-dominated programs are likely to be the slightness of employee interest and appreciation. No matter what steps are taken it will be very difficult to stimulate enthusiasm.

2. The actual decisions are likely to be unwise. The decisions tend to be one-man decisions and lack the perspective of the wide-ranging interests and attitudes of the employees themselves. More input does not guarantee better decisions, but it does increase the likelihood when the decision will affect a great many persons.

3. The emphasis of the employee program tend to be placed wherever the interests of management lie—too often in the direction of varsity-type sports.

4. Financial support will be uncertain and varied. Personnel management changes. The amount of financial and other support will depend to a considerable degree upon the interest, the ability, and the "clout" of the personnel executive assigned responsibility for employee activities.

5. There is a lack of continuity. One management director may be quite supportive of employee activities, but the next director may not be. As often as the personnel department is reorganized, there is a hiatus until clear lines of interest and authority are established. Meantime activities may flounder. Sometimes lost momentum is never recovered.

6. There is the strong likelihood of paternalism. "Paternalism" suggests a relationship between employer and employees which resembles the care and the control of a father over his

children. Paternalism in a father indicates that the father is either overprotective or domineering. Paternalistic practices may be motivated by widely different emotions. They may come about simply because the father cares tremendously for his offspring and is too timid to trust critical decisions to other hands but his. They may come about because his children represent an extension of his own ego rather than individuals with their own lives to live. They may come about because the father regards his children as his possessions, which, like other possessions, he feels free to use to his own best advantage. They may come about because he distrusts the ability of his children to make good decisions that would meet with his approval.

The paternalistic father may be overindulgent or overrestrictive. In neither case is he likely to meet with a great deal of success in rearing his children. Once it may have been different. Now, here in America, there is so much value placed upon democracy and individualism, that the prevailing mode is for teenagers to insist upon the right to develop their own values and make their own decisions.

In this social climate, paternalism on the part of an employer is almost certain to be met with disdain and at least covert resentment.

Neither fathers nor employers should give everything to their proteges nor use their largesse for control. There are employee activity clubs which seem to thrive in spite of evidence of paternalism but one wonders whether the fine facilities and extensive programs can really make up for the disdain and resentment engendered by the domineering management. We believe that without exception paternalism is deleterious, often extremely so, to American employee activity programs. Most notably, paternalism is destructive of the high employee morale which should be one of the greatest benefits of industrial recreation programs.

The line may be hard to draw but it is important that the employer be aware of the need to draw a line somewhere between paternalism and indifference. American is far removed from the age when employers were prone to consider their employees simply as tools of production. Support, without dictation, of employee recreation organizations is an indication of being in the mainstream of contemporary American social and economic thought. We do not find that view anywhere better expressed than in the statement of Henry Ford II, taken from *Top Management Speaks:* (preface to this chapter)

> There is an important and valid place for recreation activities in industry. This is by no means a paternalistic viewpoint. . . . In a direct, practical, and forthright way, it expresses the company's concern for the human well-being of its employees and its gratitude for their essential contribution.

Disadvantages of Association Independence

By the same token, employees ought also to be concerned about the welfare of the company. The independent employee association may have fewer disadvantages than the company-dominated program, but those few are very serious.

1. It violates one of the principle strengths of industrial recreation: provision for cooperative, mutually advantageous relationship. The tie between them is certainly not very close. They are more like in-laws than family members. If a crisis, financial or otherwise, should arise, the association should not expect the company to come to the rescue.

2. The close and valuable communication that is one of the benefits of a cooperative association will not likely develop.

3. The sense of identification, important to the employee's emotional and psychological well-being, will be lessened.

4. The association will be lacking the invaluable expertise which might otherwise be provided by the company. That expertise may be in management, in legal affairs, or in financial affairs.

5. Financial assistance from the company will not be forthcoming. If a company contributes financially to the employees association, it will and must, since the company management is responsible to the company board of directors, have something to say about how the funds are spent.

In many instances revenue from vending machines on company property is one of the most important sources of employee association revenue. Some persons may not consider that revenue a company contribution, but it certainly is. The company contributes the space for the vending machines, it supplies the electricity, and it foregoes the profits which would otherwise accrue to the company.

If the company provides financial support to an employees association, the minimum it should expect includes:

1. The contribution should not be wasted by unwise use.
2. The contribution should operate in such a way as to improve employee morale.
3. The contribution should benefit all employees and not a chosen few.
4. The association should not flout established company policies.
5. The organization should not bring disrepute upon the company.

All apart from financial considerations, cannons of fairness suggest that the employee association must be mindful of its influence upon the company image. A company inevitably will be held responsible in the public opinion for the activities of its associated (or dissociated) recreation association. The organization exists only by virtue of employment in the company and invaribly uses the company's name in its own name. What the association does can greatly harm the public image of the parent company. Reasonably, the company should be kept informed, and have some influence, if not an actual veto, in respect to matters crucial to the company's public relations.

Determinants of Relationships

It may as well be recognized that the determination of the relationship of company and employee association will be made by the company. Those interested in the formation or further development of employee recreation organizations may, if tactful and persuasive, influence the company management, but management will make the ultimate decision.

The size of the company will probably have a significant bearing. In small companies, perhaps under a thousand employees, the likelihood is that the company will keep a large hand in the operation of the recreation association. The natural relationship between the employees and top management will be closer. They will have more interest in each other as persons. The company public image will be more greatly affected by employee association activities.

In small companies, it usually is not feasible for the association to have much in the way of either equipment or facilities unless the company contributes significantly. Instances exist, particularly in small companies, in which the close personal involvement of the top management results in a more successful employees recreation association.

The closeness of the relationship is particularly likely if the company is the chief employer in a rather small community. Both the size of the company and the size of the community are relevant factors.

Because of the many variables, it is not easy to generalize about the optimum degree of closeness between the company and the employee association. Yet, whatever the circumstances, employees will have the greatest interest and participate more extensively if they are given a large voice in making decisions.

Unions and Industrial Recreation

Another element which keeps cropping up, and which may affect the company-association relationship, is the presence or absence of unions, or the management's attitude toward unions. This is unfortunate, but true.

It is also true that there are many instances in which a union conducts its own recreation program. Perhaps because of the increasing level of education and sophistication of union leadership there is concern about the whole life of union membership as well as with working conditions and pay. That interest we commend. However a full discussion of union-operated recreations, service, and benefit programs should be included in a book about unions. We shall not pursue it here.

This is not to say that union members should not be included in employee recreation organizations. Quite the opposite is true. Article 30 of "Principles of Industrial Recreation" includes the idea that "a special effort should be made to involve union personnel on an equal basis with management personnel as fellow employees."[4]

That this can be done quite successfully is demonstrable. There are highly successful industrial recreation programs in companies which are fully unionized, in companies which are partly unionized, and in companies which are not unionized.

The Jet Propulsion Laboratory of the California Institute of Technology has permitted its employee recreation association to be incorporated separately. Yet at the time of incorporation a "Statement of Understanding" was carefully drawn up which included the provision that the incorporated recreation club should not engage in labor negotiations, politics, or religion.

This is reminiscent of the reply of a NIRA Consultant in answer to a question about the role of unions in the formation of a new employees recreation club. That answer was:

> We, therefore do not form the organization as a coalition of management and unions. We recognize the total work force as one industrial family without reference to any outside agency or influence. We do not intend to be promanagement or prolabor, regardless of what unions we have or have not. The intent and guiding principles of the organization are to help each other, create stronger friendships, and work for the prosperity and welfare of all. In such organizations by total vote will emerge leaders from both union and nonunion ranks. However, they never discuss or reflect their biases in organizational matters. Like religion, it is best not to let it influence the organization or have any effect on its management.[5]

Although it sometimes develops that what seems theoretically desirable is not practically feasible, they do coincide in this instance. In the Cummins Engine Company of Columbus, Indiana, there are two independent unions. One is composed of the shop hourly workers; the other, of office hourly workers.

The Cummins Employees Recreation Association (CERA), incorporated since 1961, has found no problems resulting from the unions. The CERA Board is composed of ten members. Five come from the Shop Hourly group, two from the Office Hourly group, two from the Exempt group (supervisory and administrative employees), and one member represents Management. The hourly employees on the Board do not represent the unions but the employees who elect them.

The unions have at times offered their membership opportunity to participate in softball championships, pool tournaments, and other limited events, but neither union has tried to duplicate the activities or facilities available through CERA. The relationship of CERA and the union leadership has been good, and the relationship of CERA and the Company has been good.[6]

An Example at Sunstrand

A most unusual arrangement at Sunstrand Corporation at Rockford, Illinois, illustrates the bearing of the historical background in determining the form of the employee recreation organization, and it also demonstrates how union and nonunion employees share a common recreation association comfortably. At Rockford, there are unionized manufacturing plants and the large nonunionized company offices.

At Sunstrand there are two separate recreation associations. One of them, the Sunstrand Activities Recreation Association (SARA) is largely company operated. The other, the Sunstrand Employees Recreation Association (SERA), although not incorporated, is largely independent and elects its own officers and board.

In 1967, Sunstrand put up a new activities facilities building. Although designed for all divisions' employees, it soon proved inadequate and is now used primarily for the activities of the office employees. It may, however, by arrangement be used by SERA, the employee-run recreation organization.

SARA focuses on intramural programs, hobby clubs, and arranges all travel tours.

SERA has a seventeen-acre park; but makes extensive use of community facilities. The union sponsors a team in the City Bowling League, but has no recreation program as such. There is no competition for the benefits and privileges of the recreation association. There is some good-natured competition for election to offices, but both sides get elected. Some office workers and supervisory personnel come up through the manufacturing work force and are popular with both sides. The SERA Board is composed of union and nonunion employees, but in either case they represent the employees as employees and work together quite harmoniously.[7]

The Relationship between Objectives and Organization

In the Beginning

As the great majority of employee activity associations have small beginnings little or no thought may be given at that time to the ultimate objectives of the organization. The organizational structure usually is the simplest one that can meet the immediate aims of the group. Therefore many recreation organizations just "grow like Topsy." More often than not it is only after the organization has expanded its areas of concern and experienced some difficulties that the importance of objectives and organizational form are appreciated.

This is not meant to discourage the formation of an employee recreation club with very limited objectives. Getting started with almost any kind of organization is better than postponing everything until all things are possible. All that is needed is a small number of persons who have the desire to initiate at least one common activity among their fellow workers. Those persons or a steering committee, if the numbers are large, can draw up a constitution or set of bylaws. This then should be submitted to as many employees as can be interested in the idea. With its approval and the election of officers, the organization is on its way.

Nevertheless, if some thought is given in the beginning to possible future objectives, that is advantageous. The original organizational structure should be planned so that it may be expanded if new activities or services are added. Yet even though thought is given to broader objectives at the beginning, the original objectives are certain to need reexamination, and probably revision, further down the road. Serious consideration of purposes and objectives may be brought about by a proposal for a radically new departure in activities or services. It may come with the hiring of a professional director who is aware of the importance of objectives as a sort of roadmap toward progress.

The Development of Objectives

The development of meaningful objectives which will serve well as a guide for future development cannot be achieved in a single board meeting. Brainstorming in a board meeting may set the process in motion. Time and patience are required for the emergence of many ideas and for their testing through wide discussion.

It is of great importance to involve the whole membership in the setting of objectives. This cannot be done in a general membership meeting. Probably a special steering committee should be appointed. Either the board or the steering committee itself should appoint a considerable number of special committees. Every person who is willing should be placed upon one committee or the other. The steering committee can divide up the areas for consideration and develop a timetable for reports, preliminary and final. It can coordinate the work of the committees and combine the reports into a tentative document. It can then explain the provisions of the tentative document to the whole membership, and, if the reaction indicates the need, revise the document before it is submitted to the vote of the whole membership.

Unless the members of the steering committee are impressed with the importance of the project and are themselves serious and dedicated, little worthwhile may be accomplished. The individual committees will be prone to postpone their reports to the last minute and then throw together some hastily conceived ideas. In fact the individual committees need to begin work early. After discussions among themselves, they should solicit ideas from members of the organization, management, and persons who are familiar with the industrial recreation field or some aspect of it. The committee may wish to prepare a survey to be sent to all the members. If several committees plan a survey, the steering committee should combine the various surveys into one. The response will probably be better.

Some committee members should be assigned to review back issues of *Recreation Management,* if that is at all possible. Others may correspond with or visit officers or the recreation manager of organizations with outstanding reputations. All these things take time, but the end product will be worthy of the time spent. The extensive involvement of the whole membership will bring forth many good ideas. In addition it will assure that the membership will support the steps taken to achieve the objectives.

Intangible Objectives

To many persons, objectives may suggest only the aims of the organization in terms of activities and facilities. In the larger sense objectives also include such intangible goals as the development of leadership, improved communication, and so forth. The objectives should be reflected in the organizational structure, as they are at the Nationwide Insurance Activities Association of Columbus, Ohio.

One of many good examples, the Nationwide association has a five-member board of trustees elected for three years, with staggered terms. The officers of the board, elected from its membership, serve one-year, nonconsecutive terms. Thus during the three-year term the member will serve in two and probably three different capacities.

Below the board the four councils: sports, social events, cultural activities, and services. Members of the councils are elected from the different activities or subclubs. The council elects its own officers. The council chairperson, or a designated substitute, attends the meetings of the board of trustees, to provide input and carry back information.

The association member may join any activity or activities. There is then the opportunity to be elected as an officer of that subclub or activity and to serve on the appropriate council. In the council, there is the opportunity to be elected as a council officer, perhaps the chairmanship with the responsibility to represent the council at the meeting of the board of trustees.

Almost certainly those who serve as council officers become prime candidates for election to the board of trustees in future years. Thus in gradual stages, many members will gain leadership experience and leadership ability. This arrangement provides for an unusually wide opportunity for leadership identification and development, while also providing for democratic involvement at every level.

One of the objectives of the Nationwide organization is the development of leadership. That this was no accident is suggested by part of a speech made by Martha M. Daniell when she was the Nationwide Recreation Director.

> As the participant grows with assuming greater responsibilities from a member of a council group to the chairman of that council and representative to the board of trustees and finally member of the board, so does his experience in administration grow.
> On the level of the board he learns about operating a business organization in the areas of:

> 1. Developing guiding policies and keeping them updated.
> 2. Controlling funds like a modern business organization.
> 3. The responsibilities to the city, state, and federal government of a business.
> 4. The role they may play in the community.
> 5. Awareness of individual needs and growth of employees.[8]

The Development of Leadership

There are many other organizational forms that will serve the objective of developing leadership. Regardless of the form, there are three primary requirements.

1. There must be provided opportunities to try being a leader at a low level of responsibility. This provides new insight. Some persons will find the necessary attention to details distasteful. Some will find responsibility burdensome. Some will be unable to submerge their own egos sufficiently to secure the cooperation of others. Some will find they enjoy leadership.

2. There should be the opportunity to advance by easy stages to positions of increasing responsibility and challenge. Along the way there will be the need to make public presentations and to interact with small groups. If needed, the person may take a short course in public speaking, join a toastmasters club, or do some reading on the subject. With improvement comes greater confidence and increasing ability to function as a leader.

3. Finally, the organization should provide recognition for leadership well exercised. Public recognition is reward for what has been achieved and encouragement to continue.

The Benefits of Leadership Development

Leadership service is of course an asset to the employee activity association as well as a form of individual self-fullfillment. The capacity for leadership, when recognized often leads to promotions with the parent company, sometimes culminating in very high positions. Additionally leadership ability developed in employee activities may find expression in leadership services in the community at large.

Communication as an Objective

In modern times the importance of effective communication has been recognized far and wide. Among the astute, there is also recognition of the difficulty of communication. If communication is an objective of high priority, the organization should provide representation of all the different divisions of the parent company. Properly organized, this may provide good communication among the employees, between the employees and the activity association, between the employees and management, and between the activity association and management.

An excellent example of organization facilitation of communication is the Jet Propulsion Laboratories Employee Recreation Association. Obviously with communication in mind, its by-laws state that the governing council should be made up of representatives elected from each of the divisions and districts of the company. We repeat here *Section 3.*

Duties and responsibilities of JPL ERC Division Representatives:

(a) To conduct, manage, and control the affairs and business of the ERC, as a member of the Council of Division Representatives.
(b) To serve as the liaison between the ERC and their division.
(c) To see that miscellaneous flyers are distributed to all members of their division, both for the ERC and JPL, announcing activities and events such as dance, picnic, luau, movies, Christmas party, blood bank, savings bonds, charity fund drives, etc.
(d) To see that miscellaneous flyers are distributed for the subclubs, announcing their activities and events such as bowling and golf tournaments, fishing trips, club meetings, etc.
(e) To see that bulletin boards are maintained and monitored in their assigned area, keeping the boards updated, neat and orderly.
(f) To assist in maintaining a source of two-way communication between employees and JPL management through the ERC office.
(g) To provide a means of reporting news that might be of particular interest by simply contacting the ERC Office.
(h) To inform the ERC Office and Secretary of personnel changes, for the purpose of maintaining records of the number of employees in their division.
(i) Appoint someone to handle distribution of materials if they are planning to be absent from JPL for an extended period of time.

These provisions of the JPL ERC are quoted less to emphasize the mechanics than to note the emphasis placed on communication. Item (f) is significant because it suggests the need for good communication between the workers and management and that the recreation organization can assist. The ERC professional administrator, in possession of the information that is supposed to come to the ERC office, will assume communication with management as one of his important responsibilities.

Varying Organizations

We have discussed only two of the important intangible objectives as relating to organization. There are a number of others, but there is no ideal organization. It is almost certain that the kind of organization which maximizes the opportunity to achieve some objectives will not serve equally well for others. It becomes necessary to rank order the priorities and to seek an organizational structure which is most useful for the most important of the objectives. This is why the setting of objectives is so important. From where we are, it is no more possible to recommend an organizational structure for a particular association than it is wise for a doctor to prescribe for a patient he has not seen.

The Question of Incorporation

For the same reasons we refrain from making a specific recommendation in another matter at least tangentially related to organization. That matter is the issue of whether an industrial activity association should incorporate. There are experienced and knowledgeable persons in NIRA who take opposing views on the desirability of incorporation. There will be a number of factors concerned in the decision, the most important of which will be the wishes of the company management. Anyone who conceives a strong desire for the recreation association to incorporate must first of all convince the company management.

There are seven reasons given for incorporation. Some of them affect the company more than the employee association.

1. Clearer Lines of Authority

The lines of authority are clearer and more certain. This is true in most instances. The articles of incorporation and the accompanying by-laws usually make quite clear where the authority lies and what persons (positions) may exercise what part of it. The board of directors will have the power to govern, and the officers are responsible to the board. The board governs but does not own. Ultimate power, then, resides in the membership which is the collective owner. Changes in the by-laws are clearly the prerogative of the whole membership.

2. Greater Independence

In most cases the incorporated association will have more independence and enjoy more freedom from company interference than their nonincorporated counterparts. Yet complete independence is more an illusion than a reality. The activity organization will probably still be somewhat dependent upon company financial support and assistance. As a minimum, most parent companies continue to employ the professional director of activities and whatever staff the director has.

Even though the incorporated association receives no financial support from the parent company, complete independence is not possible. If nothing else, clearly the officers of the corporation, the individuals who make up the board of governors, and the members themselves all get their paychecks from the parent company.

Complete independence is impossible. Richard M. Brown, Ph.D., CIRA, recreation director for the Texins, Incorporated, and member of the NIRA Board is admittedly biased toward incorporation. Yet he acknowledges bluntly, "One way or another, the company is going to have control."[9]

3. Ownership of Property

Corporations may own and convey property. The ownership of property by an association is usually impossible. In some states the transfer of real property to a nonincorporated association is illegal. In other instances the attempt to convey real property to an association will be treated as a transfer to the individual members instead.

Whether the ability to own property is of considerable importance is a moot question. Many unincorporated recreation associations have the use of extensive and often highly developed property, though the actual title is in the name of the company. In some instances there is a written agreement that if the company should ever require the property for business purposes, the employee association will be compensated. Whether written or not, there is a clear moral obligation. It is almost inconceivable that any company would unfairly dispossess its employee association of well-developed recreation facilities. It would seem that actual title to property is more a matter of pride and ego satisfaction, which is not unimportant, however.

4. Limited Company Liability

The incorporation of the employee recreation association does relieve the company of liability, so far as civil damages are concerned, for injuries arising out of the use of the recreation facilities. It does not relieve the company of liability in respect to Workmen's Compensation, a matter which will be discussed at greater length in the next chapter.

Unquestionably, the incorporation of the employee association does somewhat reduce the company's exposure to civil damages resulting from recreation activities. Whether that advantage is great enough to offset possible disadvantages, only the company management can decide.

5. Liability of Association Officers

Once corporations officers were believed to be virtually immune. In recent times, there are many instances where damages have been lodged against the officers as well as the corporation, association, or government agency involved. Claims against the officers personally can be made when it can be established that the officer was guilty of illegal wrong doing, or when the officer failed to exercise due care to prevent malfeasance on the part of other employees.

Presumably these developments explain the availability of personal liability insurance for directors of corporations and directors of unincorporated associations. It appears that the likelihood of personal liability claims is not very high, but that the exposure is about the same for the directors of corporations or nonincorporated associations.

6. Company Income Taxes

In some instances the company may derive income tax savings by the incorporation of the employees club. If, for instance, all the revenues from vending machines belong to the employee corporation, those revenues would not have to become a part of the earnings of the parent company and be figured in the company's tax liability. On the other hand, if the company's expenditures for the employee association are greater than the vending machine revenue, there may not be any tax advantage. Each situation is somewhat different. Probably no generalization about company tax savings is possible. Almost certainly each company explores with its tax lawyers the tax consequences before it makes a decision to permit its employee association to incorporate.

7. Nonprofit Income Tax Exemption

Income tax exemption is sometimes given as a reason for the incorporation of employee activity associations. Actually nonincorporated associations may also be eligible for the tax-exempt status. There does appear to be more inducement for the incorporated association to file for the tax-exempt status and more of them do. Officers of both incorporated and nonincorporated associations would do well to explore the pros and cons of their association's filing for the tax-exempt status.

At the outset, it should be clear that tax exemption applies only to federal income taxes. It does not affect property taxes, sales taxes, licenses, or any other kind of assessments made by government agencies.

Also, although they may receive tax-exempt status for the employee association, none of the Internal Revenue Code sections under which they may be granted such exemption, permit donors to receive income tax credit for contributions to the association. The association may profit; the donors may not. Neither are membership dues tax deductible.

There are some disadvantages to the tex-exempt status. What the disadvantages are depends upon which section of the Internal Revenue Code is used. In general these disadvantages are limitations upon eligibility for membership and/or sources of revenue, amounts of revenue, kinds of business activities, and relations with the public.

Two other observations need to be made about this complex matter. (1) The present advantages and disadvantages rest upon regulations of the U.S. Internal Revenue Service. As regulations, they are subject to change any year. (2) Even with the present regulations, much is unclear and is left to the interpretation of different internal revenue officials. In 1974, and there has been no legislative change since then, the Commissioner of Internal Revenue noted that the entire area of tax-exempt organization is in a volatile state. He added: "The Internal Revenue Service relationship with exempt organizations is highly complex and is, by necessity, as diverse as the thousands of organizations seeking to establish and maintain tax-exempt status."[10]

Fortunately the avenues for exploration are much clearer than the complex subject of tax exemption. Begin with the free booklet, "How to Apply for Recognition of Exemption for an Organization," Publication 557, Department of the Treasury Internal Revenue Service. That document reflects the lack of clear and uniform standards that surrounds the area of tax exemption, but it is a good beginning. The nearest Internal Revenue Office will cheerfully answer your questions about your own specific circumstances. It will provide the necessary forms should you decide to make application. Above all, consult a qualified tax specialist before making a decision. Logically a good place to seek help is with the company tax attorney(s) or tax accountant(s).

The Process of Incorporating

If there is a decision to incorporate, the basic procedures are deceptively simply. But statutory provisions differ from state to state, and an attorney should be consulted.

The two necessary documents for incorporation are the Articles of Incorporation and the by-laws of the proposed corporation. These will be filed by the incorporators, of which the required number varies. The incorporators are required to act in good faith, deal fairly, and disclose all relevant facts.

The usual requirements for the articles of incorporation are found in Section 28 of the Model Nonprofit Corporation Act. They are:

1. The name
2. The period of duration (usually perpetual)
3. The purpose or purposes for which organized
4. Any provisions, not inconsistent with law, which the incorporators elect to set forth in

the articles of incorporation for the regulation of the internal affairs of the corporation, including any provision for the distribution of assets on dissolution or final liquidation

5. Address of its initial registered office, and the name of its initial registered agent
6. The number of directors constituting the initial board, and the names and addresses of those so serving
7. The names and addresses of each incorporator.[11]

If, for tax-exempt purposes, members wish to form an unincorporated association, they need only to agree to associate. The by-laws may constitute the formal agreement. Some states require a signed article to be filed. Conformity can almost be assured if the articles filed contain the provisions required of nonprofit corporations.

Now, having devoted so many pages, we conclude the chapter with brief discussion of less numerous but very important kinds of organizations in the area of industrial recreation.

Inter-company Recreation Associations

It is possible, as amply demonstrated, for the employees of a number of industries in one community to combine into a single large recreation association. The outstanding advantage is the greatly increased financial base and, hence, the larger and better recreation facilities. Another advantage is the greater sense of kinship of all the employees. Additionally, there tends to be greater involvement with the total community.

The disadvantages are the loss of identity with the employing company, less opportunity for intra-company communication, and less opportunity for the administrative involvement of the rank and file workers. Also lost is the intimacy of smaller associations and with it the psychological and emotional support that can come from smaller programs.

An outstanding group association is the Industrial Mutual Association of Flint, Michigan. Begun in 1901 as a benefit association of Flint vehicle manufactureres for the insurance of employees, it gradually expanded into recreation activities. Later it was combined with a YMCA sponsored activities association for all factory employees of the area.

The tremendous financial resources have made possible unusually fine recreation facilities. The IMA has a five million dollar Sports Arena, emphasizing hockey and skating. There is also a large IMA Auditorium and an auditorium annex, an excellent golf course, and a beautiful lakeside lodge.

The IMA is strong in community cooperation and shares its facilities with the community. Unusual features are a special Safetyville, a miniature city to teach youngsters traffic safety. Also for the children there is an IMA farm.

In Mexico

A much younger example of a similar pattern is the ALPHA Industrial Group Employees Association of Monterrey, Mexico. Stronger than the individual company associations might be in resources, it offers a wide variety of activities and services for the employees and their families. In addition to an Association health care facility, there is an Association store and a library. Families are involved in the Association Recreation Center in recreational and cultural activities. Employee representation in the various levels of governance is encouraged.

Other such organizations exist as the product of unique historical forces or as a result of a special set of circumstances in some communities. As excellent as IMA and ALPHA are, their kind of recreation association probably always will be more of the exception than the rule. Another and quite different kind of organization appears to be of growing significance: community industrial recreation councils.

Recreation Councils

The recreation councils are steadily increasing in numbers. They provide the opportunity for social and professional contact of recreation directors and for those personnel officers whose assignments include employee recreation. They provide a mechanism for stimulation of enthusiasm and for the exchange of ideas and the evaluation of recreational experiences. They encourage professionalism.

Equally important, the association councils provide the means by which the advantages of group travel and merchandise and commercial recreation discounts can be made available to the employees of quite small companies or company branches.

The Phoenix (Arizona) Council is an excellent example. It numbers more than seventy members. Included are banks, hospitals, saving and loan associations, state and local government agencies, grocery stores, automobile dealers, as well as manufacturing companies.

Especially strong on the West coast, the recreation association councils are spreading across the nation. Most of them owe their being to the enthusiasm of one or two professional recreation administrators. Among these are the Toledo Industrial Recreation and Employee Services (TIRES), which serves northwest Ohio and southern Michigan. Its current vice-president is a member of the recreation faculty of the University of Toledo. The closer association of employee recreation associations and institutions of higher education is a healthy trend.

Also distinct, The Oklahoma Industrial Recreation and Fitness Council was incorporated in 1977. It features strong ties between the Oklahoma State Government, the State universities, and industries and businesses of all kinds. As its name implies, its present emphasis is upon physical fitness.

The Benediction

In concluding this long chapter, we offer our blessings and a little parting advice to the newly graduated student at the beginning of a career in industrial recreation and to the personnel officer assigned as a novitiate to the company employee activity program.

We do firmly believe that "form should follow function" and that organizational form is of much importance. However, we urge you not to "throw up your hands" if you "inherit" a situation where the organizational form is undesirable and apparently not amenable to change. A great deal of change often can be effected by persistent, gentle, and tactful persuasion. Moreover, a great deal of good can be accomplished no matter what the form of the organization.

It has been said that there are three parts of wisdom. The first is to know how to change the things that need to be changed. The second is to be able to accept the things which cannot be changed. And the third is to know the difference.

Chapter 5 should therefore be of interest to you. It is about the functional aspects of organization.

Notes

1. Donald E. Hawkins. "The Formulation and Validation of Principles of Industrial Recreation in the United States." (Ann Arbor, Michigan, University Microfilms. Ed.D Dissertation, New York University, 1967), p. 439.

2. George D. Webster. *The Law of Associations: An Operating Legal Manual for Executives and Counsel* (New York: Mathew Bender and Company, Inc., 1975), 2–24. Webster is General Counsel for the American Society of Association Executives.

3. *Industry Week*. March 26, 1975, pp. 28–32.

4. Hawkins. "The Formulation and Validation of Principles of Industrial Recreation in the United States." p. 443.

5. Mel Byers, CIRA, to Nancy Fitch, Lynbrook, N.Y., October 1976.

6. Information courtesy of Steve Waltz, CIRA, Cera Manager.

7. Information courtesy of C.L. Polaski, CIRA, July 22, 1976. Mr. Polaski was Manager of the Activity Center and management representative to the SERA Board, where he served as an advisor without a vote.

8. Martha Daniell, a past President of NIRA, is now retired from Nationwide. Her speech, undated but identified, was found among material collected for this book.

9. Manuscript submitted by Dr. Brown for this book.

10. Webster. *The Law of Associations. 15–5.*

11. *Webster. 2–11.*

5

Functional Aspects of Organization

TOP MANAGEMENT SPEAKS

Management has an obligation, wherever possible, to see that employees achieve a sense of personal satisfaction from what they do.

Goodyear's early management strongly believed that there is a direct correlation between a healthy body and a person's ability not only to do his or her job, but also to enjoy fully the hours away from work. Thus, more than a half century ago, a large gymnasium was built opposite corporate headquarters in Akron. It has been in constant use ever since.

We actively promote the idea of employees taking part in sports. Our company has leagues for basketball, flag football, softball, bowling, volleyball, and golf. In recent years membership in our skiing and tennis clubs has mushroomed. Realizing that not everyone is cut out for active participation in sports, Goodyear sponsors 41 clubs that offer a varied menu of interests, ranging from chess and bridge to gourmet eating and model railroading. A 1,400-seat theater provides a fine setting for the Goodyear Musical Theatre. The 75-acre Wingfoot Lake Park is becoming one of the finest employee facilities in the country. The company sponsored its first Boy Scout troop in 1914. Today Goodyear is one of the largest industrial sponsors of scouting in the world. We also sponsor the world's largest hunting and fishing club with a membership of more than 5,000.

All of this costs money, but we consider it money well spent, because it lets our em-

C J. Pilliod

ployees know we care about them over and beyond what they produce on the job.

Recreation becomes a key factor in productivity by giving the individual status and recognition, as well as improving his or her morale. A diversified selection of activities for employees, members of their families and retirees is the fiber that joins our people into a well-knit, friendly group—on and off the job.

C.J. Pilliod
Chairman of the Board
Goodyear Tire & Rubber Company

TOP MANAGEMENT SPEAKS

Years ago, George Eastman, founder of Eastman Kodak Company, wrote: "What we do in our working hours determines what we have. What we do in our play hours determines what we are." Eastman's statement is a good expression of the value of recreation time well spent. It is also a statement that applies to all of us . . . to the young and the old, the administrator in his office as well as the man on the production line.

Each of us, of course, must decide for himself how he will use free time. No one should tell you or me what our hobbies, interests, and activities ought to be. But worthwhile interests can be fostered by the availability of programs and facilities designed to appeal. In this sense, industry seems to have a real opportunity. If athletic fields, gymnasiums, reading rooms, photo labs, hobby shops, and other facilities are made available, the men and women of a company will be far more likely to enter into and benefit from rewarding activities.

Recognizing this fact, Kodak and other companies have made such programs and facilities available to their employees.

What is more, the management of our company is convinced that the money spent to encourage healthy spare-time activities pays big dividends—both for Kodak and for the people of the company. Few of us can afford to be overly introspective about our work. We cannot isolate ourselves from the world around us, concentrating solely on what takes place in the office, the laboratory, or the plant.

In this highly complex age, industry has great need of people who are healthy, vigorous,

Walter A. Fallon

and competitively keen. Industrial progress depends, in large measure, upon the insight, the interest, and the depth of understanding that large numbers of people bring to their jobs. The man who spends his free time well cannot help but bring valuable outside experience to his job inside the company. Upon such people our future will depend. What those men and women "have" when they come to us is not nearly so important as what they "are"—and can be.

Walter A. Fallon
Chairman and Chief Executive Officer
Eastman Kodak Company

In this chapter we are concerned with such functional aspects as finances, facilities, and legal matters. Although they are all interrelated, it makes for clarity to treat them separately. This we do, and in the order mentioned.

Finances

The Costs of Industrial Recreation

It is impossible to get even a reasonably accurate figure of the costs of industrial recreation in the United States. Not the least of many obstacles is the reluctance of top management of most large organizations to discuss the company expenditures for employees recreational activities. This is probably because they feel it unwise to reveal to stockholders expenditures which are substantial in amount but which produce intangible returns. Sometimes top management lacks a firm conviction of the merits of industrial recreation.

Fortunately there are other company presidents who are strong believers in the value of industrial recreation and are willing to translate their convictions into dollars for recreation facilities and activities. Among them are Walter A. Fallon of Eastman Kodak and C.J. Pilliod of the Goodyear Tire and Rubber Company. Complete statements from both are at the head of this chapter, but we repeat portions here for emphasis.

Walter A. Fallon:

> The management of our company is convinced that the money spent to encourage healthy spare-time activities pays big dividends—both for Kodak and for the people of the company. . . .
>
> If athletic fields, gymnasiums, reading rooms, photo labs, hobby shops, and other facilities are made available, the men and women of a company will be far more likely to enter into and benefit from rewarding activities.

C.J. Pilliod:

> Goodyear's early management strongly believed that there is a direct correlation between a healthy body and a person's ability not only to do his or her job, but also to enjoy fully the hours away from work. Thus, more than a half-century ago, a large gymnasium was built opposite corporate headquarters in Akron. It has been in constant use ever since.

The Extent of Company Support

In the United States, in contrast with most other nations, the trend is toward cooperative financing. Increasingly where there are active employee associations, the association contributes or generates a substantial portion of the cost of employee activities and services. Yet there is a significant difference between different geographical regions. As a general rule, employee associations have spread most rapidly in the more industrialized states and least rapidly in states in which the economy is not dominated by industry. In North Carolina, where the industries tend to be relatively small and rather dispersed, "The one type of recreation organization which is rarely used . . . is the Employee Association."[1] Yet North Carolina is the home of a number of full-service employee recreation programs with excellent facilities and a full-time director, one example being the Olin Corporation of Pisgah Forest.

An example of a company program from a different geographical area is the Foster Grant Company, Inc., of Leominster, Massachusetts. At Foster Grant, the recreation director is directly responsible to management and he runs the show. Yet employee participation has been good, partly, at least, because the varied activities are directed by participant-elected committees and the emphasis has been upon family type recreation.[2]

For one last example of an excellent "company" program, we go still farther away to the Dominion Foundries and Steel, Ltd., of Hamilton, Ontario, Canada. The top management of DOFASCO, as it is called, has long been committed to employee recreation and began subsidizing a recreation program in 1937. In recent years its emphasis has been upon varsity type programs and family participation. Almost entirely company supported, a slight exception is made in the large number of hockey teams composed of employees children. The more than four hundred players between the ages of eight and fifteen are all "completely equipped with a suitable uniform provided jointly by the parents and the company."[3]

There are of course a great many other examples, some with recreation programs of the order of those mentioned. It is observable that there is more and more employee involvement in decision making, at least at the individual club level, in even the strictly "company" programs. And it is still our contention that, by and large, the better programs are those which are *by* as well as *for* the employees, and we see that as the trend for the future.

Although not so numerous, there are employee associations which derive all or nearly all of their financial support from the parent company. At Rockwell International Corp., the "total recreation program is financed by management."[4] Although the funds are drawn from the budgets of the different divisions, they are deposited to the separately incorporated "Rockwell Aid," which is the administrator of the recreation and welfare programs.

We believe that all companies pay at least a part of the costs of operating an employees' activities and services program. In a number of instances the company contribution is in the form of vending machine profits, which the employees often do not recognize as the contribution it really is. In many circumstances, Management also treats the vending machine profits as though they were not a company donation. Since neither the revenues nor the outgo appear on the company's books, stockholders are unaware of the financial support given to the employee activities program.

A great many companies contribute to the employees association on the basis of some kind of formula. The formula usually involves the number of employees at the site, the number engaged in the recreation program, or the amount contributed by the employees themselves.

At the Eli Lilly Company the recreation association's funds are received through the sale of activity membership cards. The Company then matches those funds two-for-one.[5] The Johnson (Wax) Mutual Benefit Associations derives funds largely from membership dues, which are matched by the Company, from the fees charged by special clubs, and some other company contributions. The Company pays for all capital purchases, and underwrites the costs of certain tournaments, the company picnic, and special events which the Company believes to be especially beneficial for public relations.

Employee Preferences

For the most part, employees feel a greater interest in programs to which they contribute. Dr. John H. Rapparlie is a corporate psychologist who has served as a consultant in many personnel studies. His research and experience have led him to believe that the employees really do not want everything provided for them. As Dr. Rapparlie related it:

Our participants have expressed preference for a program supported no more than 50% by company funding, refuting the commonly held opinion of management that employees are only interested in a "free ride" and in "give-away" programs.[6]

In part the desire of employees to provide much of their recreation funding stems from the pride of doing something for themselves. To a larger extent it may result in a desire for a greater degree of independence. They recognize the element of truth in the old saying that "Who pays the piper, calls the tunes."

Not unnaturally, the employer who contributes the funds feels entitled to control their expenditure. At Dofasco, for instance, although the various subclubs are self-governing, "the Advisory Council is made up of members of Management and Departmental Superintendents and the (company hired) Recreation Director. "Company policy is interpreted and the yearly budget is set" by that Advisory Council.[7]

At least one company recognizes vending-machine income as a contribution and accordingly controls the recreational budget. The official company manual of the McClean Trucking Company of Winston-Salem, North Carolina, is of interest in a number of respects. In the portions reproduced below, the emphases have been supplied by the authors.

> *FUNDING*—All terminals will subnit all commissions, profits, etc., from the regular vending machine programs to Winston-Salem on a regular daily cash report. *The entry on the cash report should clearly indicate that it covers profits, commissions, etc., from vending machines.* This money along with additional sums added by the company will be used in a variety of ways such as entry fees, equipment, portion of bowling sanctions, rental of facilities, and other company sponsored activities. Terminal Manager will submit to the Recreation Director, Personnel Department, Winston-Salem, with approval of District Operations Manager, Division vice-president, and Executive vice-president, Field Operations, at the beginning of each calendar year a budget for his anticipated activities and amount of funds needed for the year. . . . The Budget Committee *will approve or modify* the requested funds and advise the terminal manager of his allocation for the year.

From the Manual also we find the criteria used for the distribution of the funds. They are unusual in employing a combination of objective and subjective criteria.

> The amount of funds allocated for employees at a given terminal is determined by the following factors:
>
> 1. The number of employees at that terminal.
> 2. The interest exhibited by the employees in the various activities.
> 3. The availability of local recreation support and local recreation facilities and programs.

Income Sources Related to Programs

Another reason for the growth of employee or association contribution to the costs of industrial recreation programs is that the employees often desire a more expensive program than the company is able or willing to support alone. In 1976, Richard M. Brown, General Manager, Texins Association, conducted a personal survey of a large number of NIRA member organizations. In analyzing the results he found that those programs with facilities averaged an annual expenditure per employee about three times the average of those programs without facilities ($32.00 to $10.70).[8]

Brown also found a direct correlation between the sources of income and the amount available to the organization. The greater the number of sources, the larger was the amount available per member. His findings are represented in the table below.

Average Income from Various Sources

Method	Number of Companies	Average $/Employee/ Year
Company Contribution Only	9	$17.31
Company Contribution and Dues	9	$22.66
Vending Only—No Dues and No Company Contribution	6	$10.71
Vending + Dues + Other—No Company Contribution	5	$27.16
Company Contribution + Vending + Dues + Other	11	$30.31

Dues and Fees

Some dues and/or fees probably should be charged even if other funding sources are operative. If employees have some financial stake in the program, they tend to participate to a greater extent and to be more interested in accepting some responsibility. Hardly any program is so wealthy that it can not make good use of some member dues or fees. With the additional revenue, new activities can be added and more services provided.

Payment should be voluntary, but for those who wish to participate, a payroll deduction method is a great convenience. Payments are more regular also and budget planning is made easier when the payroll deduction method is used.

It is essential that in deciding upon a dues and/or fee structure, the association make certain that the benefits to the members are greater than could be obtained from commercial sources for the same cost.

As explained by W. Brent Arnold, Xerox employees enrolled in the physical fitness program pay $7.00 per quarter which covers registration and towel fees. The Xerox executives have a separate facility and are charged $150.00 per year which includes a personalized fitness program as well as cardiovascular tests.

At the General Dynamics Convair Recreation Association 300 employees signed a petition asking for a physical fitness facility. With a great deal of volunteer work on weekends, the facility was built and soon had a membership of 492. The annual membership fee of $50.00 may be compared to $275 charged by outside groups. All costs, including the salary of a full-time physical fitness director, are offset by the membership fees.[9]

The Northern Natural Gas Company has a $6.00 per month charge for participants which is deducted from their payroll checks. The Health Club provides men and women with gym clothing except shoes and towels.

How much can be accomplished with able leadership and enthusiastic members is illustrated by the Texins. As noted in the March, 1975 issue of *Recreation Management:*

> Over two-thirds of Texins large yearly budget for its facilities-based operations is self-generated. The Association pays actual cost or fair-market value payments to TI (Texas Instruments) for all of its employees and for the company land and buildings it uses or occupies. And yet, even

on this basis without vending profits, Texins is beginning to become more self-sufficient and relies less and less on a company contribution for its operations.

This is the more remarkable in that the Texins have "cadillac" style facilities and programs. Its dues and fees policies are therefore worth attention.

All TI employees are eligible for an automatic membership which entitles them to participate in certain special events, ticket sales, group travel, et cetera. However, participation in the use of its recreation center; the Texoma Club, a beautiful family recreation and camping park on a large lake; the Texins Golf Center; and its many activity clubs requires the payment of fees scaled to costs and benefits. The fees which may be deducted from the payroll in a lump sum or on a continuous basis, are rather complicated to detail here but are clearly explained in the detailed TI brochure from which this information is taken.

The TI employee has several choices ranging from a "Master Membership Plan" either for the employee or the entire family, to membership in the Activities Center only, and to membership in one or more of the individual activity clubs. In every case the employee derives a great deal more from Texins fees that could be obtained from commercial sources for even considerably large amounts. Consequently membership is high and much revenue is generated.

Vending Profits

Next to company contributions and dues and fees, vending is the most important source of revenue for industrial recreation programs. Sometimes it is the only source of revenue. Though vending profits are indirect company contributions, we treat the subject separately because, in most instances, neither the company nor the association chooses to consider them as the contribution they really are.

Vending profits can be and often are quite substantial. Yet vending as the major source of revenue has one notable weakness: vending revenue is keyed closely to the number of employees, so that temporary reductions of employees may play havoc with employee association budgets and plans. Some costs do go down with lessened participation but certain fixed costs associated with the ownership or operation of facilities cannot be controlled unless the facilities are closed down or liquidated.

Vending profits can be substantial but they are not automatically so. Unless the vending operation is well-managed, the profits may be small and employee dissatisfaction great. How much difference the vending operator can make is suggested in the article, "Some Vending Operators Are Better," by F.J. France.

> We replaced the vendor in one of our plants as a result of many complaints from employees. Under the new operator the complaints not only stopped but, most surprisingly, at the same time sales almost doubled, going from $445 to $817 per week. It is difficult to put our finger on all the reasons for this improvement, but some obvious factors were better products, service through more hours, better equipment, and perhaps more accurate reporting.
>
> Good operators often provide 'extras', which are in reality good business practices—personal visits to keep us informed about sales levels and customer acceptance, immediate responsiveness to suggestions about food products and equipment, internal systems of evaluating their own operations, acceptable methods of handling customer complaints and coin refunds, and providing well-trained, understanding servicemen.[10]

The importance of good administration of the vending operation is underscored by the experience of the Cummins Employees Recreation Association (C.E.R.A.) of the Cummins Engine

Company, Columbus, Indiana. Deriving a majority of its income from vending profits, C.E.R.A. built an outstanding Park Center. With the Center and increased participation, vending profits became inadequate. It was decided to put the vending operations on a more businesslike basis and to secure competitive bids for the concessions. Profits increased even more than expected and in 1972 exceeded a quarter of a million dollars.

A Practical Approach to Vending Profits

Although this is not intended to be a manual of operations, the procedures followed by C.E.R.A. seem too valuable to omit. Also with minor variations, they are probably valid for every association which wishes to use vending as a source of revenue. They are paraphrased from an article written by the then recreation director of the C.E.R.A. His advice was:

1. Establish an advisory vending committee. It might consist of the recreation board chairman, another member of the recreation committee, a representative of Management, and the director of recreation.
2. Develop a vending contract format. With the benefit of legal counsel, it should include a 30-day cancellation clause by either party and an agreement for no price change for three years except by mutual consent.
3. Prepare a bid specification sheet. This should include the requirements for bids, the approximate number of personnel to be served, the type of equipment to be used, and the brand name of products to be vended.
4. Put the contract out for bid, stipulating a time limit. When bids have been received, check with other companies for an evaluation of the vendors who have bid.
5. From the bids submitted, make a selection based upon two factors together:
   a. the reputation of the vendor for quality of products and services, and
   b. the rate of the vending commission.
6. Recommend a vendor to the employee association board and to Management. The approval of both should be necessary.[11]

The advice about vending implies a rule that should be observed about every source of employee activity income. Arrangements for raising revenue should always be advantageous to the individual members as well as the treasury of the association. The association is the collective members. If an arrangement is disadvantageous to the members, it is disadvantageous to the association, no matter what the apparent association profits may seem to be. No deal is a good deal if it sells the individual members short. Members soon quit patronizing flakey deals, but their ill will toward the association or its management goes on and on. Morality and money both support the position that the only good deals are those which benefit members and the association alike.

Varied Sources of Revenue

There seems to be almost no end to the number of ways by which imaginative leadership may secure revenue for employee activity associations. A few of the more promising are mentioned, but undoubtedly there are more and still others will develop in the future. It seems terribly trite to say that where there is a will there is a way, but that appears to be not far from the truth. The revenue sources are listed in no special order, for the practicality and usefulness of each will vary widely in different circumstances.

Inside view of Pratt and Whitney Aircraft Club Store. It has catalog ordering and display.

A view of the P & WA Club Travel and Entertainment Center. Both money making
activities operate in the same building, but they have separate professional managers.

The Association Store

The well-planned and operated association store can be a good source of revenue, can save the members money, and can be of real service by making shopping more convenient. If "service to members" be a stated objective and if the store does business only with members and their families, its profits ought not to interfere with the tax-exempt status of an organization.

There are more association stores in operation than are generally known. In some instances the company management does not want any publicity about the store lest the competition be offensive to business interests important to the company. The really "going" concerns range in size from 800 to perhaps 8,000 square feet in space occupied, and sales range from a few thousand dollars to millions of dollars.[12]

Store offerings range through a wide range—hard goods, soft goods, jewelry, health aids, beauty aids, small appliances, and sometimes, large appliances also. The largest merchandise offerings are usually found in operations which combine on-the-shelf merchandise with catalog services.

A small store can be started with only one or two salespersons and utilize whatever expertise the recreation director or some member of the association may possess. As the operation grows in size and diversity, it will be more practical and more profitable to secure the services of a full-time store manager who has had some retailing experience.

Another option is contract with a private firm to operate the company store. The latter may seem to be the simpler solution, but it still calls for a good deal of supervision from someone in the recreation association. The professional, salaried store manager is the practice followed by the very successful store of the Pratt & Whitney Aircraft Club, Inc.

Whatever the method of operation of the association store, one thing is certain: the operation of a large store should not be the personal responsibility of the director of recreation. That person has a host of other responsibilities which are of even more importance to the association.

The popularity of employee association stores depends upon a number of factors, chief of which are the quality-price relationship and the factor of convenience. The value of such a store as an employee service is demontrated by the experience of the Chase (Bank) Employees Club in New York. After the establishment of the Chase Employee Store and Pharmacy (ESP), "membership in the Chase Club for employees quickly jumped 50% to 18,000." ESP has a former J.C. Penney employee as its manager and provides full or part-time employment for sixteen others.[13]

The Pratt & Whitney Aircraft Club store has been very successful and should inspire other employee associations. It began in 1971 with one small sales desk where logo jewelry and candy were sold. A year later the operation was expanded to a retail store which offered a few fair-traded items with a very small mark-up. In 1973 a larger number of fair-traded items were offered and at a higher mark-up though still selling at less than the prices at area retail stores. In 1974, most of the fair-traded merchandise was dropped and regular merchandise was offered at a 20% mark-up, well-under the prices of the competition. The success is evidenced by the fact that in the latter year, the store profit was the third largest source of income for the P & WA Club. From its modest beginning as a single desk operation, the P & WA Club store has grown into a large and sophisticated operation with outlets in four different company locations. It is professionally managed with a computerized inventory control system utilizing Pratt & Whitney data processing equipment.[14]

The idea of an employee association store will probably have to be sold to Management but there are strong sales aids.

1. The store can be a very substantial contributor to the employee association funds. Thus it can either relieve the Company of some of the burden or it can expand the activities and services of the association.

2. It is a fringe benefit which employees see as a convenience, and a money-saver.

3. Employees stay on the premises for shopping so they are not likely to be late getting back from lunch.

Not every sales operation need be a colossus and there are a number of other employee association sales promotions which are less extensive but do render a service and raise revenue. One of these is the retail outlet for employees only, that sells only company produced items, at prices far less than the public must pay. That is good for advertising, morale, and company loyalty.

Another type of sales operation is the selling of "loyalty" items, sweat shirts for instance, which bear the name and insignia of the employee association.

Group Travel Services

Group travel services is mentioned here because that kind of venture can operate in the same facility or in conjunction with the association store. There is the sharing of facility at Pratt & Whitney, though each is managed by a different professional. Different expertise and experience is needed.

Group travel saves money because of the bulk purchases, and the employee association can receive a discount from the transportation agencies just as may any other travel service. The association travel service may also arrange for special accommodations and receive commissions from hotels and restaurants or amusement places. This extra income is at no expense to the members. It is the association's travel service's responsibility to be sure that the accommodations or other arrangements are of good quality and fairly priced.

Photographic Service

Photographic service, film and development, can be offered alone or as part of a store operation. Little space and very little investment is required. It is a definite convenience for the associate members and can return considerable income.

Coin Operated Games

Coin-operated games in an association recreation center may encourage participation and provide significant extra revenue. Just as in vending, care must be exercised in the selection of the operator. General supervision by an association member who is alert and in touch with both the players and the operator is needed.

Full Length Movies

Some employee clubs are having great success with offering full-length movies for employees and families. A NIRA associate member offers very recent first-run movies just out of the big city theaters. There are a number of other suppliers. In situations where the association members live reasonably nearby, the movies can be an excellent source of revenue. One enthusiast suggests that profit from popcorn and candy sales can pay for the cost of the film rental, so that the modest admission prices are pure profit for the association. Which should leave everybody happy!

Business with Associate Members

Included in this topic are innumerable arrangements for employee association members to receive discounts from entertainment parks, various service agencies, and merchandisers. The Texins Association has a professional buying service to sell anything from TV's to automobiles at reduced rates.

Prime movers of the Dallas Metroplex Council, Mark Armstrong, Dick Brown, and Jerre Yoder, in writing about the financial rewards and service possibilities of such fund-raising activities, do hit hard on the subject of association responsibility. They conclude: "We stress before initiating any of the erstwhile mentioned programs, make sure they are sound. . . . Our customers don't go away. So, do *your homework before considering any project.*"[15] To which we add a hearty "Amen."

Dining Rooms

The general opinion seems to be that, quite unlike vending operations, dining rooms bring little or no profit and are just not worth the trouble. The truth is that, *if well managed,* any enterprise that makes money in the business world can make money for *some* employee associations. That the dining room is no exception is evidenced by the experience of the NASA Exchange at the Johnson Space Center at Houston, Texas. According to Timothy J. Kincaid, Manager of the Recreation Facility, the club dining room profits provide approximately 20% of the total club budget. Operating under professional management, it not only serves as a daily dining room but caters about 100 special banquets annually.

Investing

The sums of money handled annually by the larger employee activity associations are more than most of us earn in a lifetime. Not all the income can be used as immediately as it is produced. Where the interval between income and outgo is a matter of months, idle money should be invested. The Cummins Engine Company Recreation Association, already mentioned in respect to its highly successful vending operation, makes good use of its idle funds. For the larger associations, as the C.E.R.A. director Stephen Waltz observes, "wise investment practices can be the source of many thousands of extra dollars per year."[16]

Recycling Programs

There is a tendency to think of recycling programs as "penny-ante" games as far as fund raising goes. Well-organized and directed as at Control Data's employees clubs, recycling can be extremely profitable. Of 27 clubs active at Control Data, 19 are active in a recycling program. Each club has a recycle chairperson who oversees the placement bins, coordinates the activities, and urges other employees to bring contributions.

The programs would not succeed to the extent they do without the cooperation and practical assistance of the facilities administration people. The facilities administration arranges for the internal collection and delivery of wastepaper to the bins or shipping docks. This is waste which the company would otherwise have to pay someone to haul off, and it is a large amount. Large amounts also come from the homes of the employees.

Not every kind of company generates the amount of wastepaper that Control Data does; but it is likely that every large business organization discards tons of wastepaper every year. Also the average household in America discards somewhere in the neighborhood of a ton of wastepaper annually. An operation could be highly successful even though it fell far below that of the Control Data clubs for which recycling produced "more than $100,000 in revenue in 1976."[17]

Other Funding Sources

The sources of funds seem almost limitless and depend largely upon the initiative and the imagination of the employee activity director and the association governing board and officers. Among other sources that have been mentioned in *Recreation Management* are pay telephones, employee gasoline pumps, and special fund raising events. There are others.

Though emphasizing the potentials, we should caution that the raising of funds should not become such a consuming interest that insufficient time and energy are left for the development and administration of programs, activities, and services.

With that reservation, we reiterate that the employee association that has a varied source of funds is more secure and probably better off than one which relies upon just one or two sources.

There are also some interesting and revealing figures relating to "facilities," the second major division of this chapter.

Facilities

There is as wide a range of recreation facilities as there is of employee activity programs. The more extensive the program; the greater the need for facilities. Though desirable, facilities are not absolutely necessary, especially in the beginning. It has been said that "it is possible to have a successful recreation program without facilities, but it is a rough way to go."

On the other hand, it must be remembered that, as will be emphasized later, the most popular aspect of virtually every employee activity program is not recreation but services. Many services can be performed without the aid of any special recreation facilities. Certainly it is far better to start a recreation program without facilities than to delay until facilities are provided. The existence of a functioning program with significant participation is perhaps the best selling point for the construction of facilities. Moreover, an organized employees activity group can, with enthusiastic and imaginative leadership, go a long way toward providing its own facilities.

Company Provided Facilities

Innumerable companies with a sincere interest in their employees provide extensive facilities for employee activities and services. It is not so much a question of the wealth or the resources of the company as it is of Management's convictions and philosophy. Some of the very largest of U.S. corporations provide practically no recreation facilities for their employees; some small companies provide a great deal.

Certainly one outstanding example is the Flick-Reedy Corporation of Bensenville, Illinois. And unquestionably the moving force is the personal dedication of its president, Frank Flick. At Flick-Reedy, with 900 employees, there are a swimming pool, meeting rooms, and basketball courts. There are also many other facilities such as a target golf course, fishing lagoons, and a trap-shooting range.

Desire and imagination are powerful tools. Flick-Reedy built a 40 by 60 foot indoor pool for $90,000; but the pool eliminated the need for a water tower with an estimated cost of $160,000. The three man-made fishing lagoons are fed by rainwater from the roof and parking lot.[18]

A view of the excellent bowling lanes of the Goodyear Employee Center at Akron, Ohio.

The Felt Products Manufacturing Company of Skokie, Illinois, spent $1 million to buy and develop a two hundred acre park for its 1,200 employees.[19]

The Olin Corporation, of Pisgah Forest, North Carolina, has most unusual facilities for its 2,800 employees. Its 375-acre Camp Straus has a six-acre lake and some 100 acres of cleared recreation area. The remaining acreage provides mountain trails for hiking and the opportunity to "get away from it all." Camp Straus has a canteen, lodge, bath house, and a host of outdoor facilities. These include eleven picnic sheds, many other tables, a children's playground, three all-weather tennis courts, a softball and baseball field, a field and target archery range, and horseshoe pits. In addition to Camp Straus, the Gun Club has a rifle range and skeet trap within two miles of the plant. The Company also makes available some eighty garden plots ranging up to one-quarter acre for employees who wish to use them.

With the combination of abundant Company support, dynamic activities leadership, and enthusiastic members, Olin has an excellent program. Small wonder that the *North Carolina Recreation and Park Review* was able to announce in August of 1970 that Olin "has received the Helms Athletic Foundation Award for the most outstanding year-round program of industrial recreation in North America" for companies in the Class "B" Division, employing under 5,000 persons. Fritz J. Merrell, CIRA, Employees Activities Director for the Olin Corporation and NIRA president for 1977–78, had directed the program for 26 years.

Self-built Facilities

Among outstanding recreation facilities are some which were built by the employees themselves. We can mention only a few of these to inspire other associations and make them realize what can be achieved where there is sufficient determination and commitment.

An excellent, and very early example of employee-built facilities is the clubhouse of the Project Employees Recreation Association (PERA) of Phoenix, Arizona. The huge Salt River Project (SPR) which generates a great deal of electricity and also provides water for irrigation and for municipalities, has existed more than sixty years. In 1950, the SPR management donated 82 acres of land for employee recreation. The SPR management also donated the materials, but the PERA members volunteered all the labor to build an outstanding clubhouse and outdoor facilities. Working under the direction of SRP engineers and architects, the employees built "a two-story, rock-faced structure, complete with game rooms, meeting halls, locker rooms and showers, exercise rooms, a lounge, hobby areas, kitchen, and administrative offices."[20] Adjacent to the clubhouse are self-built outdoor recreation facilities.

One of the best of self-built facilities is Missile Park, the primary recreation center for the General Dynamics Convair Recreation Association (CRA) in San Diego, California. CRA's finances come almost entirely from vending machine profits, and it took CRA almost five years to accumulate the funds to develop Missile Park on 27 acres owned by the Company but assigned to CRA for recreation. But while waiting for the funds, the planning proceeded.

Different departments within the company were given the responsibility for planning and developing specific parts of the park. This insured wide employee participation and spurred friendly rivalry which contributed to quality work and the earlier completion of the formidable undertaking. Even so it took almost eight years to complete one of the finest employee recreation centers anywhere. Understandably, the pride of the CRA in its Missile Park is all the greater because "all labor with the exception of plastering, plumbing, and electrical was donated by company employees."[21]

The activity center at Missile Park is an 18,000 square foot clubhouse which provides an auditorium, the CRA offices, 12 meeting rooms, and workshops. Missile Park also includes two lighted tennis courts, a health club with saunas, a lighted softball park, a 13-acre picnic area with a children's railroad, a merry-go-round, barbeque facilities and picnic tables. It also has a model Western Town with horse show facilities.

In 1962, CRA also purchased the 86.5-acre Pincecrest Park, a mountain camping facility 40 miles from San Diego. It has been developed into a popular family camping park for the use of General Dynamics employees. Pincecrest has 195 camping sites, and the CRA owns six trailers, one A-frame, and three cabins which may be rented. It has also a 25 meter swimming pool with locker rooms and rest room facilities.

We particularly like the sentences which end the article from which this information was taken: "A basic philosophy within CRA is that it is the Employees Program; therefore, it must function on a volunteer system. CRA's training is involvement, and involvement generates leadership."

Another inspiring example is Long Hollow, the lakeside family camping park of the Lockheed-Georgia Recreation Club. The Company bought 19 acres of farm land, and a water-access of approximately ten acres is leased from the U.S. Corps of Engineers for recreational purposes. Volunteer work parties were organized to develop an access road and to clear the beach area. Since then many camping work parties were organized.

Long Hollow Park has 52 first-class campsites with water, electricity, and picnic tables. There are three boat docks, two large picnic areas with individual grills and picnic tables, a centrally located playground, two restrooms with hot shower facilities, two dump stations, two open-air patio shelters, and two storage shelters. Almost all of the roadways are completed with asphalt finish.

Development still continues as members vacationing at Long Hollow do at their own expense such projects as "building concrete steps in the camping area, purchasing and transplanting trees and shrubbery, and making directional signs."[22]

As one last example, Goodyear, at Akron, Ohio, has some of the finest facilities anywhere, but much is due to the employees themselves. When the Hunting and Fishing Club wanted a lodge, the company donated the land and gave some financial help and materials but stopped there. As the top management related the story:

> Here's the important thing: Club members, most of them craftsmen in our own plants, built the lodge themselves. These fellows worked hard and got a real kick out of it—and I know they enjoy the lodge a lot more than if it had been handed to them on a platter.[23]

There are other self-built facilities and we hope there will be a great many more in the future; but, sometimes, the answer is not building but sharing.

The Goodyear Hunting and Fishing Clubhouse of Goodyear employees at Akron.
Goodyear supplied the materials; the members built the clubhouse themselves.

The thriving Goodyear Hunting and Fishing Club releases pheasants on a nearby farm.

Sharing Facilities

Another approach to the problem of expensive facilities is sharing, of which there are many kinds. One kind is the sharing of facilities by two or more companies. For instance, Delco-Remy and Guide Lamp Incorporated cooperatively built Killbuck Park. It is an excellent 212-acre recreation park with a 158-acre golf course and is used by the employees and families of both firms.

A recent development which has promise for the future is the development of industrial parks built around recreation facilities. One, mentioned in *Recreation Management* (November, 1974) is a seven and a half acre park in Anaheim, California. It has a jogging track, tennis, volleyball, and basketball courts. It also includes an attractively landscaped nine-hole putting green, a picnic area, and centrally located shower and locker rooms that include saunas. This may be an ideal solution for smaller companies and for companies which need to be located in areas where land prices preclude extensive outdoor recreation facilities.

The sharing of facilities by industry and public parks is often a good solution. Much has been accomplished along that line in Rochester, New York. In reporting the situation there, the assistant to the commissioner of parks and recreation made what we think are some excellent observations:

"Close cooperation between private industry and local government can enrich the recreational programs of both by increasing the number and variety of facilities each has available. There is no perfect formula for developing such sharing programs. A great deal depends, however, upon the initiative of the leadership involved and upon the maintenance of a sound, friendly relationship between private and public recreation professionals."[24]

The large and facilities-wealthy Industrial Mutual Association of Flint, Michigan stresses community-wide cooperation and sharing. The IMA works with the local unions, the Board of Recreation, the Recreation and Park Board, the Y.W.C.A., and the Y.M.C.A. This cooperation is easier and more efficient because several years ago all the agencies that participated in recreation formed the *Flint Recreation Commission.* Such a commission would probably be an asset to all large communities. It should certainly be considered wherever there are multiple agencies.

On-site Facilities

With the nation-wide increased interest in physical fitness more and more companies have or are planning physical fitness facilities as a part of their office facilities. Hardly any new large office buildings are going up without some space for a health laboratory. Many companies who can are converting part of existing facilities or are arranging for employee use of nearby Y.M.C.A.

The Fitness and Recreation Center at the Xerox International Training Center for Management Development. This is a side view of the double gymnasium.

Lighted tennis courts outside the ITCMD for Xerox. Xerox has excellent physical activities centers at all locations.

facilities. In a few instances where nothing more seemed practical, some companies converted cloak rooms or supply rooms into lockers for the employees who liked to jog or ride bicycles.

Among the very finest of such office building facilities are those of Xerox Corporation, a leading advocate of physical fitness programs. In the Xerox 30-story office building in the heart of downtown Rochester, New York, there is an elaborate physical fitness laboratory for executives on the 13th floor and a multi-purpose recreation/physical fitness area on the concourse level for the other employees. The building also houses an artificial ice skating rink and an 800-seat auditorium used for both business and recreational functions.[25] Xerox provides some kind of recreational facilities, usually including physical fitness laboratories, at all its several work and training sites.

Such facilities are popular with employees, and, given a choice, perhaps most employees would opt for them. Not much research has been done about employee wishes in the United States; but Dr. Robert Wanzell did extensive research in Canada. His doctoral thesis was "Determination of Attitudes of Employees and Management of Canadian Corporations toward Company Sponsored Physical Activity Facilities and Programs."[26]

Dr. Wanzell's research established that there were differences depending upon such variables as type of employer, marital status, age, sex, and other factors, but that overall there was a strong desire on the part of a large percent of Canadian workers for on-site health facilities. Even without knowing any specifics about such facilities, a considerable majority of the workers indicated a willingness to pay a regular monthly fee for the use of such facilities.

Elaborate physical fitness facilities can be quite expensive but much can be accomplished with a restricted budget. Faced with high costs and space limitations, more and more companies are turning to the capsule laboratory type fitness facility that is endorsed by the President's Council on Physical Fitness and Sports. It is compact, relatively inexpensive and provides for a balanced work-out of the cardio-respiratory system and the major muscle groups. The amount of space and apparatus required obviously is determined by such factors as the number of program participants and peak loads. It is known that thirty persons per hour can be handled comfortably in 2,000 square feet of space. It costs approximately $3,000 to equip the area with such basic items as carpeting, treadmills, stationary bicycles, a rowing machine, wall-pulley weights, an inclined board, and a rope-jumping station.

Facility Priorities

Actually the current emphasis upon building physical fitness facilities may prove to be a mistake in the long run. The problem is that the present high level of interest in such facilities may decline and the participation may pass to some other activity. Meantime such large resources may have been invested in the physical fitness laboratory, that there may not be funds for another program if the day comes when the usage does not justify the expense and the space.

If a company sees fit to create a physical fitness facility, then that is that. But, if an employees activity association is considering investing a large sum of money in its first facility, that facility should not be a single-purpose facility of any kind. There are two reasons: (1) interest in that activity may and probably will diminish in time; and (2) no single-purpose facility will serve the interest and elicit the participation of the large majority of association members. Dr. Wanzell's research of Canadian workers, concerned only with company-provided on-site physical facilities showed that no matter what the facility, considerably less than 100% of employees would participate. Of those who indicated an interest in such a facility, 54% preferred a gymnasium type facility, 36% wanted a pool, and 10% chose handball or squash courts.

A much better example of questionable single-purpose facilities are golf courses. They are expensive to build, expensive to maintain, and seldom are used by a very high percentage of employees. According to an article written by the editors of *Golf Digest,* Monsanto Chemical Company built an employee golf course at Pensacola, Florida. We do not know the original cost, but the authors of the article note that of the Company's 6,500 Pensacola employees, 760 employees belong to the company golf club, and each pays dues of only $4.16 per month, though it costs Monsanto about $120,000 per year to maintain the facility. Using these figures, Monsanto appears to be spending about $82,000 of Company funds per year to maintain a facility used by less than 12% of its employees. That, presumably, is all aside from original cost and depreciation.[27]

We know that one robin does not make a spring and that the figures quoted prove nothing. They do, however, remind us that, as we already knew, golf is highly popular with a sector of the American population but that it is a low priority for blue-collar workers and clerical persons. But let us not stray from the original point that single-purpose facilities ought not to be highest on the priority list of employee associations.

The most used and most important employee association facility is the club house or activity center. It can provide for many different interests. It is the hub from which the program may operate most efficiently. It should provide administrative offices and be the center of communication. It provides a general member meeting place. It is practical to design a large meeting room which can be quickly and easily subdivided into two or more smaller assembly rooms. Since meetings are more often successful if food is served, an attempt should be made to incorporate a kitchen area in the activity center.

Ideally the activity center may include a physical fitness area, a gymnasium, craft rooms, dressing rooms, and an indoor swimming pool. The desirability of the latter depends greatly upon the climate. So does the appeal of indoor tennis courts.

If funds are quite inadequate for the envisioned recreation center, perhaps thought may be given to building a large shell and developing the separate parts as the need indicates and the finances permit. If even that is not possible, plan a smaller building in such a way that additions will be feasible later.

Yet here, again, we must avoid prescriptions. As important as the activity center is, some circumstances may dictate building other recreation facilities first. The company may already provide meeting rooms, an auditorium, association offices, or perhaps some of the other central facilities needed for the successful operation of an employee recreation club. In that case, it may well be that certain individual facilties, as a softball field complete with lights and bleachers, a swimming pool, or some other single-purpose facility may be sensible. Yet that is not too often the case, and it must be remembered that, taking the whole membership into account, a rather samll percentage of it will avail itself of a single-use facility. The more diverse the offerings, the greater the interest and the larger the participation that may be expected.

Facility Planning

The first step in a decision to build any facility should be a review of the association's objectives and long-range planning. The next step, if called for, is the creation of a facility planning committee. Its members should represent the employees of all work areas or divisions. The recreation director should certainly be involved, perhaps as the chairman or executive secretary. But, if the project is to be a great success, the recreation director should not attempt to function as a generalissimo.

When the planning committee has had time to organize and agree among themselves what they really want, it is time enough to consult an architect. From the beginning and throughout the undertaking, the planning committee should communicate with the whole membership. The whole membership cannot make detail decisions but it can react to planning committee reports and so influence ongoing decisions.

It is the business of the planning committee to inform the chosen architect about their objectives for the proposed facility, what kinds of activities will be going on, and how many persons need to be accomodated in each area. The planning committee must also develop the priority of their objectives and so inform the architect. It may well be that not all of them can be achieved.

It is not the business of the planning committee to layout the facility nor to draw a sketch of what they want. That is a job for the artistic talents and the expertise of the architect. Having considered such aspects as topography, soils, drainage, utilities, other natural elements, access, and environmental impact and so forth, the architect then should present a tentative plan and drawings to the planning committee. When the architect and the planning committee are in full agreement, the architect is finally ready to prepare the working drawings.

Sometimes architects develop very strong biases of their own and are inclined to substitute their judgment for the expressed wishes of the committee. Sometimes the planning committee tries to do the work of the architect. There has to be as much discussion and as many meetings as are necessary to arrive at a mutually agreeable plan. Hardly anything is more important than that the planning committee and the architects perform the roles for which each is best equipped. That may sound very simple, but, in practice, it becomes much more difficult.

Plan to Expand

Our parting word on the subject of facilities is "think big." It may be difficult, and it may require foregoing for the time being some of the amenities desired, but planners should anticipate growth. Membership interest and participion often expands terrifically because of new facilities. The Firestone Tire Employee Clubhouse in Akron, Ohio, renovated and expanded the indoor recreation and exercise facilities. The result was a 68% increase in participation in recreation programs.[28]

And, now, having advised you to "think big," we will try to "think brief" for the last section of this chapter, Legal Considerations.

Legal Considerations

To own property or to render service is to incur liability. The intent of the owner or the provider is of little moment before the law. We live in an age in which the prevailing sentiment seems to be that one is stupid not to get all that is possible. The spirit of the times is aptly expressed by an editorial in the *New York Times,* 15, January, 1976:

> This is an era when many Americans are more concerned with their rights than with their responsibilities and also a time when little premium is put upon patience or accomodation to less than ideal situations.

Legal liability is a greater hazard than in former times and it is a problem in recreation as in every other activity of life. Employee associations, incorporated or not, need to take what steps they can to protect against the hazards of civil suits.

As discussed in the previous chapter, companies may reduce their exposure to civil damages by encouraging their employees associations to incorporate. When that has been done, the incorporated association needs to carry insurance protection of its own.

Even though the employee association be incorporated, the company still may be a partner to a civil damage suit if the company owns the property upon which the injury occurs. On the other hand, the Company can ill afford to give to the incorporated employee association property which can never revert to the company and which someday may be used for purposes other than that for which it was given. The most frequent answer is leasing. The 3M Company has provided its employees association the 433-acre Tartan Park, including a championship golf course, in St. Paul, Minnesota. The property is leased to the employees club for $1 per year.

The Origin of Civil Claims

Generally employees are not nearly so inclined to bring civil damage suits as are guests or visitors. The experience of one company president, who is an industrial recreation enthusiast, is rather typical:

> We have had only one incident of that type and that came from allowing a church group to use our facilities. That caused us to change our policy concerning outside groups. We now require them to carry their own liability insurance.[29]

Any enployee subclubs which engage in hazardous activities should be required to carry special liability insurance. This is more economical than for the association to get the special coverage for all its members. An excellent example is a Ski Club. Insofar as civil damages are concerned Paul E. Copella, Insurance Administrator of the United States Ski Association believes that the Company, if not actually sponsoring the skiing trip, is in little danger of related damage suits. He offers this explanation:

> It is the ski club and not the sponsoring company that is named in these suits. Ordinarily, a written agreement firmly establishes the separation of the club from the firm where the club's members work. Although the company's name is used, liability connections are nonexistent.[30]

Workmen's Compensation

At the beginning of the workmen's compensation programs a half century ago, compensable injuries must have occurred in the performance of assigned work. Since the beginning, and especially in recent years, the necessary connection between the injury and the employment has grown more and more tenuous.

In Canada

It may well be that the process of expanding the grounds for workmen's compensation has not proceeded quite so far in Canada as in the United States. We judge so on the basis of opinions expressed by the Director of Legal Services, Department of National (Canadian) Health and Welfare, as recently as 1974. While each case tends to be unique in some sense, compensation tends to revolve around the question as to whether the employer had done what "a prudent and reasonable man would do." The courts must have a field day with that criterion.

As a general rule, however compensability rested upon the following specific circumstances, all of which would make the company liable:

1. whether the injury occurred on the premises of the employer
2. whether it occurred in the course of action taken in response to instructions from the employer
3. whether it occurred in the process of doing something for the benefit of the employer
4. whether it occurred in the course of using equipment or materials supplied by the employer
5. whether it occurred in the course of receiving payment or other consideration from the employer

6. whether the risk to which the employee was exposed was the same as the risk to which he is exposed in the normal course of production.[31]

In the United States

In the United States the courts have adopted such broad definitions of "work-related" that it is difficult to make any generalization with confidence. It is important subject for, as the amount and frequency of awards have increased, workmen's compensation has become an increasingly significant cost for employers. The implications may be especially relevant to physical fitness programs; but they apply also to all other programs, and, most particularly to those which take place on the employer's premises.

Most litigation over workmen's compensation concerns the extent of coverage offered. Forty-two states have utilized the formula "arising out of and in the course of employment." The other states have enacted legislation roughly comparable to the more widespread formula. The words "arising out of" refer to causal factors—is there a definite link between the way the injury was suffered and the work mission? "Course of employment" refers to time, place, and circumstances of the accident.[32]

If an employer requires participation in a sport or makes participation a prerequisite of employment, any injury will definitely come under workmen's compensation. That is clear enough but, as always, there are the gray areas. What, for instance, would be the court's decision if an injured softball player had been urged (coerced) by his immediate supervisor to play on the division team?

Among the tests applied by the courts are these:

1. Is the league sponsored by employers, or by employee associations? If the latter is found to be the case, the claim will be denied unless the employee can show some other form of employer connection.
2. What is the purpose of the competition? If the employer is merely attempting to raise morale of the employees, the connection may not be sufficient.
3. Did the event occur during work hours or off duty and where did it take place? If the game was played on the employer's property during work hours, the prospects for compensation are strong.

In one case the athletic association that ran the softball league was nominally independent. An employee injury was denied compensation but appealed the case. The appellate court decided for the employee because:

1. the activities were on the premises of the employer;
2. the employer gave substantial financial support;
3. the employer exercised considerable control over the association; and,
4. the employer derived substantial advertising benefits.

Significantly, compensation has been denied where the employees formed the league, but permitted where the employer promoted the league, encouraged employee participation, and employed a person as a supervisor for athletic events.

The Ohio Supreme Court held that an injury sustained at a company picnic is covered by workmen's compensation. A Maryland Court of Appeals went a little further when it held that permanent injuries sustained in touch football or at picnics are compensable *if encouraged or authorized* by the Company.

Perhaps the most recent extension by judicial interpretation, while astonishing, might have been anticipated. An employee at an office Christmas Party, which he was not obliged to attend, became inebriated, fell out of a window, and received a workmen's compensation award. Truly it seems that, no matter what the legal jargon, we are headed in the direction that the only substantial question will be: "Would the injury have happened if the claimant had not been employed by the Company?"

The question has been asked whether written and published disclaimers by employers might relieve them of the burden of workmen's compensation awards in recreational activities. One company attorney has suggested the following statement:

> "In this activity, which is optional on the part of the employee, the Management will consider that the activity is NOT 'arising out of, and in the course of employment'—and as such any illness or injuries arising out of such activities *will not* be covered by Workman Compensation. If the employee opts to participate, it will be with the full knowledge and agreement that the activity is NOT covered by the Company's Workmen Compensation Insurance."[33]

Attorneys with whom we have discussed such disclaimers have generally had two reactions.

1. The disclaimer would probably persuade some employees not to submit claims under those circumstances.
2. If the employer does choose to file a claim, the disclaimer will have little or no bearing upon the outcome.

Practical Suggestions

We seriously doubt that there are any solutions to this problem so long as social values are as they are today. We are impressed with two practical suggestions to alleviate the problem. Both were made by Mel Byers, CIRA, NIRA Consultant. The first is from "Keynotes," V. 4, No. 4.

> Now that most everyone (including industry) is under government surveillance concerning safety, it might save time, money, and embarrassment if the recreation director meets with his industrial safety director to discuss the following:
>
> 1. Injuries occurring to employees and nonemployees at activities sponsored by the association or company.
> 2. Procedures to follow concerning accidents arising from a sponsored activity.
> 3. Review of existing recreational equipment, checking for faulty or damaged pieces, and the possible injury that may result if used.
> 4. Review of safety precaution measures and their enforcements involving athletic, social and educational games, leagues, events, or programs.
> 5. Review of the areas used for recreation in light of safety precautions, maintenance of regulations, and insurance.
> 6. Review of current insurance coverage and procedures required when filing a claim.

The other general suggestion, as remembered and not verbatim, is simply that compensation claims arising out of recreational activities will be kept at a minimum if the activity chairperson reacts to every injury in a practical, humane, generous, and *personal* fashion. Such reaction should be encouraged and endorsed by the association and by the Company.

Some specific suggestions to activity chairpersons are:

1. Carry accident report forms and fill them out as early as practical. Both the recreation director and Management need to be given full details while they are still fresh in mind.
2. Immediately after every accident, take all steps possible to relieve the victim's pain and suffering, and the attendant concerns and inconveniences.
   a. Provide immediate first aid.
   b. Transport the injured to a physician's offce or the emergency room of a hospital.
   c. Assure all concerned that medical expenses will be taken care of.
   d. Personally inform the family.
   e. Provide family transportation if needed.
   f. Provide crutches or other prosthetic devices needed.
   g. Provide someone to stay with the injured's children if that will be of assistance.
   h. Send flowers.

In every possible way show personal concern for the injured and the immediate family. The good will thus generated will pay off in fewer claims and in better employee relations in the long run.

Legal Services

In closing this chapter we skip briefly to the area of legal services. We have suggested, and we still believe, that in most circumstances, employee associations should avail themselves of the services of Company attorneys. Yet that is not always the wisest course, particularly when incorporated associations become so large that their legal services are extensive. Sometimes, also, there may be a question of whose interests are best being served.

There may be other employee associations who do or should follow the practice of the Cummins Employee Recreation Association (C.E.R.A.) in going outside the Company for most of its legal work. Originally C.E.R.A. depended entirely upon the company attorneys, but, in time, it felt the need for more autonomy. The C.E.R.A. does not retain an attorney on a regular basis but consults with an attorney as the situation indicates. On occasion it seeks legal counsel outside the community. It does, however, still utilize the Company's legal department in reference to items which are part of the files of that department.

This is a matter that requires careful and considered judgment. Regardless of the association's wishes, it may be that Company attorneys will be too busy with Company matters to give immediate or close attention to the legal matters of the employees association. We would urge that the Company's leal counsel and management be kept informed about significant legal issues affecting the employee recreation association. A good and wholesome relationship can hardly be maintained otherwise.

Moving On

Having devoted two chapters to organizational aspects of industrial recreation, we are now ready to move on to what might be called programmatical aspects, specifically, scheduling, activities, and services. Perhaps the last names should have been first for, as we shall see, it commands the most interest for the largest number of members.

Notes

1. Hubert Henderson. "Industrial Recreation in North Carolina." Recreation Division of North Carolina Department of Natural and Economic Resources, n.d.

2. "Employee Recreation at Foster Grant Company, Inc.: A Program Geared to the Family." *R.M.,* January/February, 1973, p. 19.

3. "Recreation at Dofasco." *R.M.,* August, 1972, pp. 16–18.

4. Robert G. Dyment. "Why Not Include Recreational Facilities at Your Office Complex?" *R.M.,* October 1975, p. 29.

5. "Constant Growth at Eli Lilly." *R.M.,* April 1972, p. 4.

6 "Industrial Recreation—An Overview by an Industrial Psychologist." *R.M.,* January/February, 1972, p. 22.

7. "Recreation at Dofasco." *R.M.,* August, 1972, p. 17.

8. Manuscript written for this book by Richard M. Brown, CIRA, Ph.D. Brown became president of NIRA in 1978.

9. "Recreation Programming at General Dynamics." *R.M.,* October, 1973, p. 28.

10. This article, undated and otherwise unidentified, has included in material supplied by NIRA members for this book.

11. Charles W. Wilt, Jr. "Vending as a Source of Revenue for Employee Recreation Programs." *R.M.,* June/July, 1972, pp. 5–6.

12. "The Company Store Revised." *The New York Times,* 22 June 1975.

13. Ibid.

14. Von Conterno, CIRA, and Rich Dowdall. "What's in a Store?" *R.M.,* April, 1975, p. 36.

15. "Employee Services—What Your Employees Want Is What They Should Get." *R.M.,* August, 1975, pp. 8–9.

16. " 'Rolling' Over Your Money—How to Invest and Spend Wisely." *R.M.,* August, 1975, p. 7.

17. "There's Big Money in Those Recycle Bins." *R.M.,* July 1977, p. 29. Reprinted from *Contact—For Control Data People.*

18. Frank Flick. "The Untapped Potential: Industrial Recreation." Published by NIRA, n.d.

19. *Chicago Sun Times,* 21 December 1975.

20. Gary McCormick, CIRA. "NIRA Will Grow through Service." *R.M.,* August 1972, pp. 6–7. McCormick was NIRA president for 1972–73 and former manager of the PERA Club.

21. "Recreation Programming at General Dynamics." *R.M.,* October 1973, pp. 4, 28.

22. Roy McClure. "Making It on Their Own: Lockheed-Georgia Employees Build Their Own Recreational Facilities." *R.M.,* November 1975, pp. 16–18. McClure, CIRA, Past president of NIRA, is Manager of Recreation for Lockheed-Georgia.

23. Charles J. Pilliod, Jr. "Goodyear—'Something for Everyone' " *R.M.,* August, 1975, pp. 39–42. Pilliod, Chairman of the Board, Goodyear Tire & Rubber Company, was named 1975 NIRA Employer of the Year. The article is his acceptance speech.

24. Robert Dispenza. "Sharing Facilities." *R.M.,* November 1975, p. 55.

25. "Why Not Include Recreational Facilities at Your Office Complex?" *R.M.,* October 1975, p. 27. Reprinted from *Area Development Magazine,* March 1975.

26. The University of Alberta, 1974.

27. An article reprinted from the March 1974 issue of *Golf Digest.*

28. "Firestone Expands Recreation Facilities." *R.M.,* May/June 1976, p. 4.

29. Taped interview. Earl Groves, former president, Groves Thread Company, to Theodore Wilson.

30. Paul B. Copello. "Liability and the Company Skier." *R.M.,* January/February 1973, p. 10.

31. C.T. Mullane. "The Legal Implications of Instituting Employee Physical Fitness Programs." Proceedings of the National Conference on Employee Physical Fitness, National Health and Welfare. Ottawa, Canada.

32. This paragraph is based on manuscript material submitted by Arthur L. Conrad, J.D., CIRA, vice-president for Employee and Public Relations, Flick-Reedy Corporation. We are indebted to him for most of the research on this subject.

33. Ibid.

6
Recreation Programing

TOP MANAGEMENT SPEAKS

Industrial recreation programs represent a sound investment in a company's future. And that is how they are regarded at United Technologies.

Recreation—whether in group activities or in individual pursuits—offers something to all of us. It can expand our interests, teach us new skills, strengthen our bodies, give us a new appreciation of our environment, or simply help us unwind from some of the pressures of everyday life.

My business experience has convinced me that people who regularly engage in some form of off-duty recreation generally perform better on the job. Another important reason for organized industrial recreation programs is their contribution to employee morale and team spirit. They instill a sense of belonging.

We have given close attention for many years to a balanced program of recreational opportunities at each of our plant locations. The programs now include activities not only for active employees, but also for their family members and for retired employees.

I'm proud that United Technologies' Pratt & Whitney Aircraft Club, which observed its 40th anniversary in 1975, received the National Industrial Recreation Associa-

Harry J. Gray

tion's 1975 award for the best industrial recreation program among companies with more than 10,000 employees.

Harry J. Gray
Chairman and President
United Technologies Corporation

TOP MANAGEMENT SPEAKS

Employers have been hearing a good deal recently about how they can make jobs more varied and interesting for their personnel. Perhaps some improvements can be made to make our jobs more fulfilling and thereby enrich our lives, but there are limits to this approach.

A much more unlimited area for improvement lies in our leisure hours. With holidays and vacation time, employees now average only about 20% of each year on the job. How they spend that other 80% of their time is obviously of vital importance to their happiness and self-esteem.

The 17th century poet George Herbert told us: "He hath no leisure who useth it not." Doubtless a reincarnated Herbert would maintain that the worker who spends most of his weekend sprawled in front of his television set has wasted his leisure time. That's a position which can be argued. However, the Monday morning blahs are probably much more likely to strike the workers who did nothing in particular on the weekend than whose who went golfing, sailing, camping, fishing, skiing, or model airplane flying.

The latter have changed their environments and attitudes on their own time. The former, having failed to use their leisure time to "get out of the rut," may return to the job feeling deeply mired.

That's one of the reasons why our company sponsors literally hundreds of teams, clubs, and special interest groups among its

Sanford N. McDonnell

personnel. In this way, our people who would like to try a new activity, hobby or sport can find ready-made groups that are anxious to welcome them into membership and to teach them the ropes.

We're going to continue our efforts to insure that our personnel have the best opportunities to make good use of their leisure.

Sanford N. McDonnell
President and Chief Executive Officer
McDonnel Douglas Corporation

The major thrusts of this chapter will be program development and recreational activities. They are handled separately only for emphasis and ease of organization.

Program Development

Many philosophies should be utilized in determining the need for program activities. Activities should be developed with a purpose in mind and there should be a need for each activity. Need may be determined through employee interest, the professional opinion of the program director, and the goals and objectives of the total program.

The program director should set realistic guidelines for program development and continuation. Some of the questions that a director might ask are: How many teams must I have to make an effective league? What is the maximum number of persons that can be enrolled in this activity? What is the minimum number of persons we must have to make this activity worth the effort of administering it? Is this activity worth the expenditure? Many factors such as those mentioned are used in establishing the value of an activity. Many of the answers for the pertinent questions can be found in the goals and objectives.[1]

Programs should flow out of objectives. Yet, if there arises a real need for activities or services which do not fall within the framework of the organizational objectives, the objectives ought, at least, be reexamined. We are pleased to observe that objectives are in the minds of some of the leaders of industrial recreation. For that reason we reproduce here the objectives which Richard A. Riley, Chairman and Chief Executive Officer, gave for the development of programs at the Firestone Tire and Rubber Company. He began by expressing a willingness to sponsor any hobby or interest in which sufficient numbers were concerned and continued from there:

> The willingness to begin new activities, I might say, is one of six objectives which guide the operation of our employee activities program. The other objectives are:
> To organize, sponsor, and develop programs which are of positive value to individual employees, their families, the Company and the community
> To offer activities which are meaningful for the participants and which serve to strengthen bonds among and beween the company and employees
> To encourage all employees to take an active part in the activities of their choice and to guarantee to every employee an equal opportunity to participate in any activities that are offered
> To provide the best available facilities, equipment, and administrative personnel for employee activity programs consistent with budget funds
> And, finally to provide honor and recognition for achievements in employee activities and thereby develop self-esteem and pride in personal, team, and company accomplishments.[2]

In addition to being geared to the organizational purposes, employee activities and services need also to mesh with social change. This idea has been well-expressed by John T. Fey in *Top Management Speaks* [See preface to chapter 10].

> The concept of employee recreation changes through the years. This is a natural development since recreation programs are initiated and administered by employees whose attitudes and interests, in turn, are ever changing.

At least part of the change of interests of the employees has come about because of the changing character of the work force. We note only a few of these which have special significance for programs of industrial recreation. A study by the U.S. Census Bureau reveals that "between 1950 and 1975 the number of women in the labor force nearly doubled. . . . "Especially significant for program development, "the labor force participation of women with children under age 6 has tripled from 11% in 1950 to about 37% in 1975."[3]

Other relevant changes in the labor force were noted by Daniel J. Haughton, 1973 NIRA "Employer of the Year." In his acceptance speech, he mentioned the larger number of young adults in the workforce, more white collar workers, and increasing numbers of minority groups in better-paying jobs. He related the younger workers to new subclubs at Lockheed: "astronomy, science exchange, bagpipe bands, scuba diving, judo, soaring, and a lot more."[4]

At least as important as the work force changes are the broad social changes, of which two have had particular effect upon recreation programing: interest in the family, and interest in the outdoors. Perhaps both belong in the category of "endangered species."

The recreation movement is away from the single interest phase of program activity to the involvement of the total family. Now that there is roughly one divorce for every two marriages performed, we are increasingly conscious of the need to preserve the family. Partly out of the feeling that families who play together stay together, but for other reasons also, there is a growing desire of families to participate together in recreational activities.

Some of the same factors influence the new emphasis upon nature and the outdoors. As the cities become blighted and expand into already crowded suburbs, there seems to be a game of leapfrog. People leapfrog from the city cores to the suburbs and from the suburbs to more rural areas. But industry follows people. Already along the coasts and in the sunbelt, the expanding rings of large cities run into each other and create the megalopolis: a huge, densely populated, almost unbroken urbanized area. Some people are becoming environmentalists and associating with others of the same mind. Others are content to enjoy the outdoors while they can; hence there is a new emphasis upon walking for pleasure, backpacking, touring, and camping. These changes simply add to the constant need for variety.

The Need for Variety

Early in the history of industrial recreation, the emphasis was upon athletics and amateur sports. Later the thrust moved to intramural sports—the competition of teams with the company. Some of the old guard still cling to the idea that industrial recreation is athletics. For the new breed of professional directors, the rule of thumb is "something for everyone." The intent is to involve the greatest number possible. No recreation or activity will appeal to even half the employees. The more activities there are, the greater is the likelihood that some will appeal to everyone. Sanford N. McDonnell represents this point of view in his statement which prefaces this chapter. He quotes th poet George Herbert to the effect that "He hath no leisure who useth it not." That is why his company, McDonnell Douglas Corporation, "sponsors literally hundreds of teams, clubs, and special interest groups among its personnel." Smaller companies can hardly have hundreds of activities but each should have as much variety as is feasible.

Some concern has been expressed that the offering of a large variety of programs will cause participation in each to suffer. This is true only if the programs are very similar. If the activities are sufficiently diversified, there need be no concern about the total number of activities offered.

There is too much preoccupation with numbers anyway. The number of participants is far from the only measure of the value of an activity. A particular activity may appeal to only fifteen or twenty members but appeal to them very much. Except for that activity, some might not participate at all.

Skating for the employees and families, Battelle Columbus Laboratories at Columbus, Ohio.

Fencing at the Battelle Columbus Laboratories for members of the recreation association.

Success of an event rarely should be measured solely by attendance. Some of the sorriest failures have been attended by large groups. Smaller groups have some advantages, such as more flexibility, comradeship, and active participation.

There must also be variety in the events that continue from year to year if, indeed, they are long to continue. The constantly repeated successful event will ultimately lead to staleness and failure. No matter how successful the year before, each chairperson should try to give an event some different twist than it had before.

Programing New Activities

Good programing is one of the basic needs to keep an organization alive and interesting. The art of programing requires talent, skills, time, and effort. Often a little research regarding a proposed activity will pay off well. Where has it been tried? Where has it been effective? How was it developed? These are only some of the questions.

It behooves the program director to be alert to the trends of the times, reading widely, keeping in touch with other recreation programs, reading *Recreation Management,* obviously, but also reading other magazines, and being conscious of public interests in television programs. The director also needs to know what is going on in the schools. Because of a broader concept of education today, schools are providing an ever greater variety of extracurricular activities. Industrial recreation can build upon the interests created in the schools and colleges. Their graduates will be the employees of tomorrow.

Determining the needs of an employee organization requires skillful direction and working harmoniously with the membership at large. Usually the program director is in a better position to evaluate the recreation and services needs than anyone else. Yet the good director never assumes a dictator's role or discourages employee injection of ideas, suggestions, or effort. The administrator should be one who can relate well with employees, developing a position of trust and confidence without upstaging anyone. Curiously, the best role of the director is not to direct, but to stimulate, to introduce, and to educate.

The membership is a prolific source of ideas for new programs. A common policy is that if fifteen or more members want a new activity or club it will be organized. Yet that is not always the best policy. There are a number of other factors to be considered. Also, as everyone knows, a very few highly enthusiastic advocates can get thirty members to sign a petition for an activity in which less than half will participate even at the outset and which will be down to perhaps a half dozen in a short time.

Activity interest surveys are also common approaches, but that method has its dangers. Not everyone who means well can design or interpret a survey successfully. Surveying is a learned skill, and, unless the leader is educated in the technical skills of surveying, it might be well to involve someone who is—perhaps a knowledgeable professor from the nearest university.

Another problem with surveys is that people tend to express interest most often only in the things they already know about or have experienced. We can hardly quarrel with those who say, "Give them what they want," but that is an oversimplified solution. The strongest contrary view is that which was expressed in 1964 by James C. Charlesworth, then president of The American Academy of Political and Social Sciences:

> Patronizing is to be avoided but the [recreation] department should never be happy to "give the people what they want." Most people do not know what they want when it comes to developing their intellects, their personalities, and their bodies. They should be encouraged or compelled to learn to like things they presently dislike.[5]

The middle ground, as suggested by Stephen Waltz, CIRA, is: "Give them what they want as well as what they will learn to want." Since program directors are professional persons, they should be capable of moving people into new areas of recreation and into new experiences.

We would urge administrators to be receptive to new ideas from whatever the source, and not to be held back by fear of failure. If something does not fail once in a while, not enough new ideas are being tried. Nothing tried until proven, nothing failed: these are the marks of stagnating programs.

Program Priorities

Not all activities suggested, nor even all of those desirable, can be introduced at one time, if at all. There have to be priorities. One priority is program balance. We highlight this thought from the statement of Harry J. Gray, of *Top Management Speaks,* at the beginning of this chapter.

> We [at United Technology Corporation] have given close attention for many years to a balanced program of recreational opportunities at each of our plant locations. The programs now include activities not only for active employees, but also for their family members and for retired employees.

Another high priority is programs which will attract beginners and not just those persons already interested and skilled. As one analyzes the program to offer, it would be well to take into consideration the things that everyone can do, those that most persons can do, those that a fair percentage can do, then those that fewer ones can do.

The recreation director must also be conscious about the relationship between cost and benefit. Is the proposed program the best expenditure for the Company or the Association? Is there another program which can be provided for the same or less expenditure and which will provide more value received for the participants? When, if ever, may the program become self-supporting? There are other equally relevant questions.

The recreation director should also be concerned about what programs are offered and available elsewhere in the community. When industrial programs are coordinated with community programs unnecessary duplication of facilities and programing may be avoided. Ideally industrial recreation facilities and activities of the community should complement each other. Yet sometimes duplication may be desirable because of the extent of demand, the location of facilities, scheduling requirements, or other reasons.

Patience in Development

Patience is an outstanding virtue in every administrative position. After the initial suggestion for a new activity, it takes a considerable amount of time to ascertain the extent of the interest, whether that interest can be sustained, and whether it can be expanded. Organization and promotion also take time. A very sensible procedure is illustrated by the way one recreation director explored and then organized a ski club.

A few employees approached the director about help in getting a ski club organized. An informal survey suggested that a considerable number of others were interested. An organizational meeting was held to explore further the extent of interest. The attendance and enthusiasm were good, so a committee of five were chosen to work with the director on the project.

Monthly meetings were subsequently held to discuss and vote upon policies and procedures for the club. Six months after the initial meeting, the constitution was adopted by the charter members and an election of officers was held. Within a year the "Hotdoggers Ski Club" had almost 100 members.[6]

Promotion

Much depends upon how effectively a new activity or service is promoted. Poor promotion can lead to the failure of great ideas. That is one reason for extensive membership involvement. The committee members that become totally involved and enthusiastic about their "baby" will reinforce each other's enthusiasm, and that enthusiasm will spread throughout the organization. Enthusiasm is the yeast of success.

The enthusiastic committee also needs the aid of an enthusiastic recreation director who can contribute essential expertise. Effective promotion depends upon communication, one of the hallmarks of the good recreation director. The director should make sure that posters and flyers are prepared and that they are concise, accurate, understandable, and eye-catching. All available media of communication should be used: bulletin boards, the Company or Association newsletter, informed announcements at all kinds of employees meetings. Department heads should certainly be informed, for they are in an excellent position to pass the word along. Meantime the director must be alert about the reaction to the proposed program.

All this applies to all aspects of the recreation, activities, and services program, hence, obviously to recreational activities.

Recreational Activities

This recreational activities section is organized into five parts: (1) athletics, (2) outdoor recreation, (3) cultural activities, (4) travel, and (5) hobbies. These do not exist in watertight

Womens basketball.

Womens softball.

Mens slowpitch softball.

Mens fastpitch softball.

A few of the team sports of the Battelle Columbus Laboratory employees. Women are quite involved also.

compartments. But this arrangement has made the writing easier for us, and we hope it will make the reading easier for you. The treatment of the different activities is brief. We have to be content with a book and not an encyclopedia. A book could be written about each activity.

Team Sports

Every recreation club or association of even a hundred members ought to attempt one or more team sports. If there are not enough interested members to form a league, form a team and join a league. If there are not enough for a team, get with another small organization and have a joint team. Team sports do have important virtues in addition to those which accrue from all types of physical activities. Team sports are great for generating camaraderie, which will improve morale and the individual's sense of belonging and identification, with the club and with the company. Team sports increase group spirit and cohesiveness even for those members who do not play. They are also excellent avenues for the discovery and development of leadership.

Much of the time, the most enthusiastic performers, often the team captains and leaders, are younger employees and, hence, not in high supervisory or administrative positions. Thus their work-day position is suddenly reversed and they are in a management position with the need to accept responsibility, make decisions, solve problems, resolve disputes, and impart enthusiasm.

It is important that the sport be open and inviting to everyone, not dominated by the super-athletes. The very highly skilled must be divided among the teams or they must have a separate league. Otherwise the average or inept will never venture into the activity.

It is also worth remembering that not everyone wants to be a star or even a performer. There are always members who enjoy the opportunity to participate as team managers, score-keepers, and in other nonplaying capacities. Their recruitment is important to the team and to the individuals themselves. It is also important for the director of recreation, who will thus have the time to do the other things which need to be done.

Soccer.

Basketball.

Baseball.

Pratt and Whitney Aircraft Club athletes in action.

Volleyball. Mixed teams.

Administrators of employee programs should try to avoid becoming involved in the nitty gritty routine of time-consuming details such as team organization—who plays what position?—keeping score, arranging schedules, keeping track of the equipment, coaching or officiating. To do so will lead to the neglect of other responsibilities. No one can do a good job if spread too thin.

Women in Team Sports

The participation of women in sports has increased dramatically in the past few years. Particularly it has increased in those associations or clubs which take care to encourage female sports activity. Most of the women in industry today graduated from schools and colleges when the participation of women in physical sports was far less common than today. The world of athletics was not considered very ladylike. The female teams received little administrative support in the way of qualified coaches, adequate supplies and equipment, and fair access to the school's athletic facilities. Consequently, many of the women in industry have little or no athletic background, and they often need special encouragement to begin an athletic activity. Because of inexperience, the women will often need more individual and team coaching. Once started, they need recognition—on a par with the men—and continued encouragement.

In at least one instance, partly because of the initial light turn out of women employees, the decision was made to invite the participation of employees' wives. The result was enlarged participation and enthusiasm with a larger and better functioning league. At the Johnson Wax Company, a separate bowling league was formed for the wives.

At the Johnson Wax Company headquarters in Racine, there are about 500 women employees. Unquestionably a significant number of that 500 do not participate in athletic sports because of age or medical conditions, yet more than 100 women participated in the softball league in the summer of 1974. That is astonishing participation.

Again, involvement is the key. The recreation director who wishes to introduce female athletics or a new team sport, needs to begin with a small committee of enthusiastic women. With their input and enthusiastic promotion success seems almost guaranteed.[7]

Girls *Women's* softball is very popular at Goodyear.

Two volleyball games simultaneously at Xerox. Note women also on teams.

Though the development of women's teams and leagues is desirable, it is also well to remember that the line between the sexes is no longer mountain-high nor indelible. Mixed teams also have much to offer. There is no reason, for instance, not to involve men and women on the same bowling teams—the handicapping will take care of individual differences. A fixed ratio of male-female players per team would also make the same idea practical for volleyball and other team sports. Tennis tournaments already involve mixed doubles, and the practice could be extended further than it now has. It would be good for family involvement and would certainly improve participation of both sexes. The same ideas apply to the nonteam sports.

Nonteam Sports

As important as team sports are, there are many persons who prefer to play only with a buddy or in small groups. They also must be served. Of these kinds of sport activities, the biggies today seem to be golf and tennis, with the latter appearing to gain momentum in the backstretch. Tennis is now becoming a year-round sport as facilities are built to accommodate indoor courts. Platform tennis, which requires much less space, is rapidly gaining its enthusiasts. Behind these, and too numerous to mention in their entirety, come a whole host of others, from the increasingly

Horseshoes.

Handball.

Pistol shooting.

Rifle Club.

Physical activities at the Pratt and Whitney Aircraft Club.

popular handball and squash to old-time favorites such as billiards and horseshoe pitching. All of these may be made more exciting, and new adherents recruited, by company and inter-company tournaments. NIRA has a number of tournaments, and, for almost every sport, there are regional and national tournaments conducted by some association. Among NIRA's tournaments are golf, trapshooting, rifle-pistal, skeet shooting, and fishing; which lead us naturally to outdoor activities.

Outdoor Activities

The last few years have seen a virtual explosion of interest and participation in outdoor activities. Camping is high on the list. Many recreation associations have made provision for camping parks for their membership. Others are planning this really important addition. Even those companies in the most concentrated population areas are usually no more than a few hours drive from an attractive area which, often by the labor of the membership, may be turned into an inviting camping park. So many persons have their own recreational vehicles (RVs), distance is somewhat less important than the attractiveness of the site. Nearby water is very important.

There seems to be an idea that people go camping to get away from it all and to avoid the crowd, but far the larger number of people camp in developed parks with other campers quite nearby. Notably also, the typical RV, trailer or motor home, has all the facilities and conveniences of modern living. Look around a camping park and you can count more TV antennas than fishing rods.

Actually, most people do not go camping for the sake of camping, but for a variety of other reasons. According to a 1972 survey, the breakdown of reasons for camping was as shown in the following table.

Reasons for Camping[8]

Activity	Percent Listing the Activity
sightseeing	66
fishing	64
swimming	55
hiking	49
boating	48
hunting	28
backpacking	19

For the parents, camping represents the opportunity to give the children a treat and, at the same time, get them out from underfoot. During the daylight hours, the busiest spot in the camp probably will be the children's playgrounds. Impromptu softball games are numerous. Also—less inhibited than adults—children love to socialize with newfound friends.

To a great extent, choice of outdoor activity is related to age. The teenagers and young adults tend to go for skiing, snowmobiling, and cycling. The noisy snowmobiles and motorcycles are unpopular with most adults and are associated with danger. Yet these activities can be made reasonably safe with adequate safety education and proper equipment. The safety education and information about equipment might become a service function of the recreation club.

Bicycles are also mushrooming in popularity. Campers with children and some without, frequently take the bicycles along. Moreover, bicycling at home for fun, exercise, or transportation continues a long upward trend. A bicycle club is sure to attract a fair share of interest. This, in spite of the fact that bicycling is rated as more dangerous than football. The latest craze for the young, though, is for mopeds (motorized bicycling), still more dangerous. Only time will tell whether that interest is fad or lasting fashion.

RV Tours

If your club does not have a campground, or even if it does, an interesting special event is an organized group RV tour. With a little boost from the recreation director, some RV enthusiasts will probably form a committee and organize a tour during the vacation season. This is an excellent means of fostering a sense of closeness among the employees of larger firms. It may not appeal to a high percentage of members, but it may be vastly enjoyed by those who do opt for it.

Perhaps recreations directors ought not eat or sleep; for there are so very many activities which need "only a little boost from the director." Not only do these include team and nonteam sports and outdoor activities, but there are also the many cultural activities.

Cultural Activities

So what are cultural activities? Ignoring the dictionary, we include exposure to or participation in whatever appeals primarily to our esthetic senses or intellects, or, that which is characterized by education, imagination, or creativity. Once "cultural activities" were conceived by the working class as "high hat!" But no more!

What is a cultural activity? Well, if it is not athletic, or "outdoorsy," gambling, or sexually oriented, it is probably cultural. It includes not only the opera, ballet, and classical art and music, but also musical comedy, rock music, and belly dancing. Add book reviews, travel, the study and discussion of world affairs, and anything else about which you would like to be better informed. Above all, it includes whatever you desire to make for the sheer pleasure of it, whether it be an exotic culinary delight, a hand-carved pipe, or a model plane.

In the cultural area, as in all other aspects of industrial recreation, it is important that there be a wide variety of offerings—a cultural smorgasbord.

The cultural area is the example par excellence where the able recreation administrator will endeavor occasionally to spark an activity or event for which many persons have not yet developed an appetite. Let the circumstances and the situation provide the idea, but do something!

Wide as is our definition of culture, we are ourselves committed to the thought that the best of "culture" is that which is not merely faddish, but that which will endure and stand the test of time. Whatever your philosophy, do accept the idea that man does not live by bread alone. Whether your company is large or small, be ashamed not to sponsor some activities of a cultural nature.

Not so many employee associations can, like Goodyear, have a 60-piece band playing to over 50,000 people per year, nor a symphonic orchestra, as does Phillips Oil Company; but there is no employee club anywhere in which there are not some members who enjoy singing or acting or playing some musical instrument. You can have your chorus, small music group, or theatre club. Given the little boost from the director, your club can, like the Eli Lilly Club, experience the thrill of the club-produced musical or drama.[9]

Some things do require a good deal more than a "little" boost from the director, and one of these is group travel.

Travel Tours

Torn between including travel as "activities" or "services," we chose the former because travel is of such great recreational value to so many people. We choose to give travel tours considerable attention. In 1972 it was called the fastest growth area of recreation club activity. It is given great prominence by some recreation associations and much emphasis in *Recreation Management*. We believe that travel touring should and will continue expanding.

Finally, we confess to a nagging suspicion that group travel service is not being performed well, if at all, by the director and/or staff of some recreation associations. The success or failure—you had better believe there are enough of the latter—depends to a very large extent upon the program administrator. The administrator may truly say that the failure of a particular trip was due to the tour operator, but how was the destination selected and how was the tour operator chosen?

Travel tours are much to be desired. Travel is culturally broadening. An overseas travel trip will also enhance the ego and self-confidence of working people who earlier had considered foreign travel as the prerogative of the wealthy and powerful. Group travel develops ties between fellow travelers which continue long after the travel tour is over. It brings people from different departments and different work levels into closer association and friendship. That is good for individual and group morale and excellent for communication.

Group travel is not a terrific undertaking. Perhaps it is the expected but unnecessary complexities which cause some administrators to steer away from it. There are a few basic ideas which are worth remembering. Since in group travel, as in everything else, the approach should be something for everyone, there are a few outstanding basics.

1. Group *vacation* travel (trips of a week or longer duration) should be planned well in advance. Desirably there should be an association annual travel plan. This does not rule out some additional travel because of the development of unforeseen circumstances or special interests. The need for annual planning is that:
    a. Employees want to plan their vacations ahead of time. If you want them to join a travel tour, plan the tour far enough in advance that they can book it into their vacation plans.

The Battelle Laboratories employee chorus.

A marvelous theatre for the Goodyear cultural performances.

let yourself **GO** with XEROX RECREATION ASSN.
MEMBERS, FAMILIES & FRIENDS

8 Days-7 Nights
3 Exciting Cities $469⁰⁰

mexico

Mexico City

Taxco

Acapulco

Departing from Rochester - November 26, 1975

Promotion for a Xerox employee travel trip to Mexico.

      b. Extensive travel trips, even though in large groups, is relatively expensive, and individuals often need to take the expense into account well-ahead of time. If the employer is willing, payroll deductions spread over several months, will certainly ease the pain and encourage travel participation.

      c. It takes considerable time to prepare for group travel, for both the recreation association and the traveler.

2. In most cases the travel destination should be established and well-known vacation places. This will appeal to the largest number of employees. Only a relatively small number will be likely to have the sophistication and experience to motivate them to prefer an out-of-the-way and unknown vacation site.

      The usual and well-known vacation areas are not as great a handicap to the experienced traveler as might be expected. Count on it, even though your group travels to Spain, the majority of the travelers will spend most of the time sunning on the beach and in nearby sight seeing, plus taking the few sight-seeing tours included with the trip and of little expense. Yet the more sophisticated travelers will form small groups, rent a car, and tour the country highways and byways, gaining experiences that could not be offered and might not be desired by the majority of the travel group.

3. Choose a travel destination that offers a variety of activities for the widest appeal. That will call for fewer extras, all of which cost money and appeal to only a portion of the group. Moreover the on-site variety of activities will instill confidence in the inexperienced traveler who does not feel able to do much self-planning or exploration.

4. Keep the cost down. Some ways are: the inclusion of fewer "extras," getting the largest possible number of travelers, and limiting the length of the stay.

      There are other reasons also for keeping travel tours relatively brief. A shorter trip will leave the employee with vacation days left for other activities desired. Also, perhaps because of the cost-benefit ratio, most employees take spouses but not children. Parents usually do not wish to be away from their children for an extended time. For shorter trips, there is less problem in making arrangements for the care of the children, and the cost of their care will be less.

      The preference is for less expensive trips. The NIRA Travel Survey of 1976 revealed that 76% of the group travel tours taken by member organizations fell into what was identified as the "moderate costs" or "economy" classes. That leaves only 24% for the "deluxe" tours.

Competitive Bidding

      An important way to keep costs down and benefits up is to invite competitive bidding by travel tour agencies. In the 1977 Travel Survey [*R.M.*, February 1977] 58% of the respondents stated that they selected tour packages on the basis of competitive bidding, but only a little over 17% of the entire number sent a prepared bid form to tour suppliers. In all likelihood, real competitive bidding is impossible without the use of bid forms. Informal discussions without specifications is far from competitive bidding and will not usually produce the same advantages.

      Bidding is a real problem without the involvement of a knowledgable and experienced adviser to prepare the bid form. After receipt of the bids, the decision should not always be in favor of the low bidder. It is extremely important to check out the tour bidders. The local better business bureau is only the first step. Visit the promoter's office and observe whether it seems to be a going concern or a fly-by-night operation. By all means secure a list of the names of other organizations which have used the same tour services. The users can testify about the quality of the service and the satisfaction felt.

Once the tour is under way, it is out of the hands of the recreation club and in the hands of the tour agent. If that agent, or the agency staff, is inefficient, incapable, or careless, all kinds of snafus are likely to ensure, and many of the travelers will come back saying "never again."

Group Travel Responsibilities

Except for the conducting of the tour, most responsibility falls on the recreation administrator or staff. Fortunately, however, it need not be, and usually is not, the director or staff who puts the travel package together. For that service, unless, as in a very few instances, the recreation association has its own travel service, the director ought to turn to tour professionals. With some suggestions from the director, a tour professional will happily put together a package for consideration. This does not eliminate competitive bidding. The preparer of the tour package will have the opportunity to become acquainted and to demonstrate "know-how," but that is the extent of any bidding advantage. The recreation association or travel club should be quite careful not to accept any tour package which, by its nature, reduces the likelihood of bids by several tour agencies.

Apparently in the majority of cases, the decision about the travel destination is made alone by the recreation director or a Company administrator. We are still for involvement and would favor the formation of a travel advisory committee. The committee could help in the decisions about destinations and in the evaluation of bids. It could also be of great service in building enthusiasm. The committee should be chosen carefully and ought not be made up exclusively of experienced travelers. Input is also desirable from the uniniated.

The committee should be advisory only, because the actual details must be handled by an administrator with a telephone, secretarial services, time to make the necessary contacts, and a growing background of experience.

One of the chief responsibilities of the administrator is the promotion of the proposed trip. This is best done when there is annual planning. Initial announcements of the calendar for next year's travel tours should be out before the end of the year. Every avenue of communication should be used. A preliminary meeting regarding the planned tours could be held early. Later there should be pretravel meetings about two to three weeks before each tour take-off.

The recreation director and staff must instill confidence. It is not enough to refer questions to the tour operator. The association staff should be briefed well-enough that they can answer most of the questions that come up. Their knowledge will build confidence. They must make a point of trying to make members feel that foreign travel is not all that difficult. There are millions of adult Americans who do not know the first thing about how to get a passport, and a large percentage of them are bashful about asking. So tell them!

If a non-English speaking destination is chosen, a great idea is to arrange for classes in conversational Spanish, French, or whatever. Very often you may find a member of the company who can provide such instruction. If not, you can turn to the International Institute or to a nearby high school or college.

Of course not all travel is the vacation tour type, and the one-day or weekend tours are also good service and morale builders.

Short Tours

Tours have some of the same benefits of vacation travel and much of the advice about travel fits the tours also. It is probable that the recreation directors who have the most successful vacation travel programs are also the ones who have the most successful short tour programs. They both require commitment, enthusiasm, and planning.

Short tours demand less advance notice, but the earlier the notice the better in terms of tour participation. They also require vigorous promotion. Tours may be built around special events— a major sports event or an annual festival—or about special interests such as skiing, golfing, or the theatre. Travelling and touring are both fascinating hobbies for many persons, but there are now other hobbies which will bear brief discussion.

A ski trip for Pratt and Whitney Aircraft Club members.   A P & WA Club excursion to Yankee Stadium.

Short tours are also fun and morale builders.

Hobby Clubs

Hobbies, which many employees have and others will develop, can be of real value to the Company and the employee association or club. Some of these benefits are:

1. Higher morale. Through satisfying a need or the expression of creativity, it assists in selffulfillment and satisfaction with existing circumstances.
2. Communication. A company vice-president and a new apprentice may find a common interest in model railroads. Members come from all levels and all work areas. This fosters a sense of closeness and improves the lines of communication upward, downward, and horizontally.
3. Hobby shows. Hobby clubs lead naturally to company hobby shows that can be great for community relations.
4. Special values. In some special circumstances, hobbies have a particular utility for the Company or the recreation association. The Kodak Park Activity Association provides

Hobby shows are good for public relations, a reward for the hobbyists, and a stimulus for more participants. The picture is of a Goodyear Hobby Show in the Company gym.

an example of a hobby club promoting a company product. At KPAA the Camera Club is the largest such organization in the world with approximately 22,000 members. In addition to facilities for developing film, making prints, and enlarging pictures, the Club has several areas for classroom use and for the showing of slides and club made movies.[10]

The Goodyear Rose Club beautifies the Company grounds. The Xerox Photography Club provides support for the Xerox Recreation Newsletter.

Hobbies and hobby clubs also have benefits for the members:

1. Preparation for retirement. There is little use in retiring unless a person has found something interesting to do.
2. Increased socialization.
3. Family closeness. Often hobbies are of general interest to the whole family and include a certain amount of family participation.
4. Financial benefits. Many hobbies provide extra money. These become especially important after retirement.
5. Gifts. Hobbies are often the source of gifts more appreciated than those store-bought.
6. Recognition. Craftsmanship is recognized and provides ego satisfaction.
7. Esthetic fulfillment.

Starting Hobby Clubs

Many hobby clubs are not overly expensive and most can become virtually self-supporting. They do need encouragement, publicity, and facilities. There should be a central hobby committee. The key, as always, is involvement. From a handful of hobbyists, a committee can be appointed. After the hobby section is well-established, the central committee should probably be elected from the various hobby clubs or congregation of clubs.

The function of the central hobby committee would be to assist the recreation manager, or, even, to take charge of some of the following aspects:

1. Ongoing functions
2. Publicity
3. Coordination in respect to budgeting, facilities, and the scheduling of meetings
4. The handling of hobby shows.

Much work must be done to make a hobby show a success. Among the more important factors are publicity, the encouragement of participation, the registration and safeguarding of entires, arrangements for judges, the securing of prizes, and the awarding of prizes. Desirably, the prizes should be awarded by the highest ranking executive available. This will make the winners feel even more pleased and it will be helpful in gaining publicity.

Varieties of Hobbies

There is a tremendous variety of hobbies. Nothing else in the recreation club can so well exemplify the idea of something for everyone. Without the slightest attempt to be all-inclusive, we list some hobbies by categories. We do so merely to illustrate the breadth of the field and to provide some suggestions. Some hobbies fit into more than one category, so we had to be arbitrary sometimes. The classifications probably are not the best in the world either. We can only hope that no ardent hobby enthusiast is offended.

Artistry	**Domestic Arts**
photography	flowering arranging
ceramics	candle making
drawing	cake decorating
painting	dressmaking
sculpturing	lampshade covering
picture framing	gourmet cooking
glass blowing	wine tasting
vocal music	embroidering
instrumental music	crocheting
home movies	bottle cutting
poetry	decoupage
	macrame
	batik

Chiefly Indoor Activities	**Workshop Arts**
physical fitness	woodworking
dancing	upholstery

acrobatics
weight lifting
baton twirling
modeling
bowling
billiards
darts

Chiefly Outdoor Activities

skiing
swimming
scuba diving
ice skating
roller skating
motorcycling
horseback riding
camping
hunting
fishing
boating
skeet shooting
archery
snowmobiling
hang gliding
bicycling
hiking
trap shooting
target shooting
flower growing
gardening

Model Building and Operation

railroads
planes
boats
autos

metal working
carving
leather working
auto mechanics
auto restoring

Intellectual and Spiritual

investment
religious
ESP
transcendental meditation
transactional analysis
social service
educational
toastmasters
amateur radio
stereo and sound
indoor travel
book reviews
great ideas

Collections

guns
coins
dolls
antlers
stamps
antiques
beer cans
chinaware
paintings
sculptures
driftwood
semi-precious rocks
comic books

Unquestionably there are dozens, perhaps hundreds, of other hobbies. Some hobbies attract relatively few adherents. Others are so popular that they are organized nationally and have national shows and contests. All are appropriate for industrial recreation associations.

Physical fitness was listed as a hobby; but to many of its followers, it is far more than that. To some persons, physical fitness is recreation, but to the larger number, it probably is not. It is a very important activity, and we have had a hard time deciding where to place it. Finally we decided to place it with services, and we give it extensive treatment at the beginning of the next chapter.

Notes

1. The first two paragraphs of this chapter were taken almost verbatim from manuscript submitted by Stephen Waltz, CIRA, Cummins Engine Company, Columbus, Indiana. We are indebted to him also for other ideas included in this section.
2. "Employee Activities: A Human Investment." *R.M.,* August 1977, p. 26. Riley was NIRA "Employer of the Year" in 1977. This article is taken from his remarks in accepting that award.
3. *The Toledo Blade,* 27 April 1976.
4. "Recreation and the World of Work." *R.M.,* August 1973, pp. 23–25. Haughton is also represented in "Top Management Speaks." Preface to Chapter 10.
5. "Viewpoint in Our Time." *Recreation: Pertinent Readings.* Jay B. Nash, ed., William C. Brown Company Publishers, Dubuque, Iowa, 1965, p. 165.
6. Dale Prell. "Increasing Participation: How to Catch Employee Interest and Keep It." *R.M.,* October 1975, pp. 20–21.
7. Susan Stindle. "Women in Sports: The Female Employee in Industrial Recreation Programs." *R.M.,* March 1975, pp. 38–40.
8. Sandy Greenblatt. "The Fastest Growing Activities." *R.M.,* August 1972, pp. 12–14.
9. Donna Zink, Sheryl Huber, and Eleanor M. Fairchild. "Directing Employee Choral, Social, and Theatrical Events." *R.M.,* October 1975, pp. 22–24. See also, same issue, Roseanna M. Albini, "Planning a Company-sponsored Dinner—Theatre Outing," pp. 16–18.
10. KPAA brochure, 20 November 1975. Courtesy of Kirk T. Compton, CIRA, Eastman Kodak Company.

7

Physical Fitness and Employee Services

TOP MANAGEMENT SPEAKS

Employees don't show up on their company's balance sheet, but they are probably their employer's most important asset. As such, they should have the opportunity to develop and grow within their working environment as well as in their own homes, particularly in this age when technological advancement has done much to depersonalize jobs and the interrelationships of those who perform them. This is why, it seems to me recreational programs sponsored by the employer for the employee can contribute meaningfully to a personally rewarding spirit of community within a business organization.

At the New York Stock Exchange, we believe that our recreational, cultural, and group travel programs enhance morale and contribute to better relations among our employees. They provide a means also by which employees can share with one another some of their leisure hours as well as their working hours.

In the Exchange bowling league, on the softball field and the basketball court, across bridge tables and chess boards, and in karate classes, employees who might never know one another except by job titles become acquainted as individuals. In group travel arrangements and in dinner and theater parties, their families can become involved as well.

James J. Needham

These and other activities improve the quality of the working environment, we feel, and though the rewards may not be calculable on a balance sheet, there is no doubt that they are there.

James J. Needham
*(1967) Chairman and Chief Executive
Officer
The New York Stock Exchange,
Incorporated*

Although the company benefits of physical fitness programs was discussed in chapter 3, we discuss physical fitness here chiefly from the standpoint of the meaning and benefits to the individual. The subject of physical fitness is quite important, and the added emphasis gained by some slight repetition is warranted.

We judge its importance from its special benefits, from the amount of interest being displayed in it, from the amount of money being spent for it, and from the attention it receives in *Recreation Management*. The *Toledo Blade* in an article datelined from New York, 10 October 1975, quoted "a spokesman for the National Industrial Recreation Association" to the effect that "physical fitness programs have become the 'hot item' for corporations." Not much has changed since except that what began as a program for the top executive men has spread to all levels of employees, including women.

The Meaning of Fitness

Some persons derive pleasure from the activity itself—as distinguished from the satisfaction from the end result—but for most persons engaging in a physical fitness program is not recreation. If it is well done, it is more work than play for most persons. The brochure, "Physical Fitness in Business and Industry," by the President's Council of Physical Fitness and Sports, reads: "It will take at least three to six months of hard work to reach an acceptable fitness level."

Physical fitness is worth the effort! It is much more than the absence of illness. As the President's Council defines it:

> Physical fitness is the ability to carry out daily tasks with vigor and alertness, without undue fatigue, and with ample energy to enjoy leisure time pursuits and to meet unforeseen emergencies.

Physical fitness is not a prize waiting at the end of some particular exercise lane. Rather it is a total package of sensible practices reflecting moderation. Kishor S. Ambe, M.D., Ph.D., and Director of the Health Enhancement Institute, says that there is one nearly universal prescription: do not smoke, keep your weight down, eat more sensibly, exercise vigorously and regularly, maintain emotional composure, and avoid conditions that produce either physical exhaustion or acute worry and anxiety.[1]

Assuming the other ingredients that Dr. Ambe mentions, there is no unanimity about what constitutes a good physical fitness program. There is general agreement that fitness includes (1) muscular strength, (2) muscular endurance, and (3) circulatory endurance. Beyond the three basic ingredients there are four other components that meet with varying degrees of agreement:

1. muscular power, the ability to release maximum muscular force in the shortest time
2. agility, speed in changing body positions or in changing directions
3. speed, the rapidity with which successive movements can be performed, and
4. flexibility, the range of movement in a joint or sequence of joints.

The President's Council adds and emphasizes one other dimension, cardiovascular fitness. In "Physical Fitness in Business and Industry," the Council explains that dimension.

> When a skeletal muscle is exercised it grows stronger. Similarly, when the heart is exercised it grows stronger. . . . With the heart, improvement is determined by such standards as a reduced resting pulse rate, a lower pulse rate when certain work loads are imposed on it, and a faster rate of recovery when the load is removed.

Cardiovascular fitness is an area often overlooked in physical fitness programs. . . . But the truth is, no exercise program which ignores cardiovascular health is truly adequate

The Benefits of Fitness

The initial and perhaps still the greatest motivator of corporation interest in physical fitness is concern over the incidence of fatal heart attacks to able, young executives, often in their thirties or forties. Whether physical fitness programs will significantly lower the incidence of such heart attacks is controversial. While one may reason that physical fitness programs should have that result, there is not yet any widely accepted scientific evidence that any kind of exercise will lessen the incidence of heart attacks or prolong life.

Skeptical physicians assert that the larger number of heart attacks occur during or immediately after exercise and that is true whether or not the individual is conditioned to such exercise. Unquestionably fitness exercise programs will strengthen the heart muscles; will they prevent the sudden clot that cuts off blood to the heart, or a large area of it, called a thrombosis? This is significant because of "the estimated 647,180 deaths in 1976 from Ischemic heart disease [insufficient blood to the heart muscles] accounted for 89.5% of all deaths attributed to diseases of the heart."[2]

Fortunately, heart and cerebrovascular diseases have been on a slight decline for some time. Cerebrovascular diseases peaked in the 1920s, and mortality from heart diseases peaked in 1963, and both have declined since.[3] This is the more remarkable in view of the fact that the life span is increasing. Ultimately, of course, we all will die, and heart disease remains the leading cause of death. What is most significant is that the deaths from heart disease are occurring at increasingly advanced ages. There is a smaller percent of the 35–50 year age group suddenly dropping dead of heart attacks.

Considering the length of the trend downward in vascular and heart diseases and the still relatively small percentage of Americans regularly involved in physical fitness programs, it is not realistic to attribute the trend to physical fitness programs. Yet, we must emphasize that the absence of positive proof that physical fitness programs reduce the incidence of heart attacks does not negate the possibility that they do. Most professionals of long experience with physical fitness programs are convinced that they do prolong lives. Moreover, the actuarial statistics have persuaded some insurance companies to offer lower group insurance rates to industrial firms which have established physical fitness programs.

Yet whether or not they prolong life, there are plenty of compelling reasons for physical fitness programs. There is not the slightest doubt that a good physical fitness exercise program combined with sensible living habits will result in persons who not only feel better, but have the ability to work and play harder and longer with less fatigue. In addition, a large number of participants report that they have less tension, sleep better, and generally enjoy life more.

Probably one of the greatest of benefits is a large decrease in the amount and severity of back trouble. Very few, if an persons, are born with perfectly formed spines. With the sedentary life which is common today, most persons find that the muscles which hold the spine in correct alignment lose their strength and tone with age. The consequences range from the dull pain, usually in the lumbar region, to excruciating and disabling pain as the nerves radiating from the back are pinched by slipped discs. The President's Council's brochure, "Physical Fitness in Business and Industry," reports that the National Safety Council estimates that "backache costs

American industry one billion dollars annually in lost goods and service, and another $225 million in workmen's compensation." The *U.S. News and World Report,* 14 January 1974, observed that "backaches in 1974 are expected to be responsible for more lost work time than the common cold." And this says nothing about the unnecessary agony and inconveniences suffered by millions of individuals beginning at about the age of thirty or a few years later.

Good muscle tone can help to relieve fatigue and pain in the back, the shoulders, the neck, or any other part of the body. There is some reason to believe that it will also reduce the incidence of hernia, since the abdominal muscles will be strengthened.

Other hidden costs of physical degeneration which physical fitness can overcome include: lack of enthusiasm, a desire to call it a day two or three hours before quitting time, boredom with routine, an obsession with minor aches and pains, and general lassitude which often leads to staying home both from work and recreation without a very good reason. It is a safe generalization to say that the physically fit will be more enthusiastic, work harder, play harder, accomplish more, and enjoy life more. It is certainly worth the necessary expenditure of money, time, and effort. Those recreation associations which provide for physical fitness programs contribute both to the company and their own members.

Methods of Physical Fitness Programs

Aerobics

Of all approaches to physical fitness, probably the most technical and the one requiring the most knowledge and supervision is aerobic exercise. William DeCarlo, CIRA, defines that system thus: "Aerobics is a system of exercises stimulating the heart, blood, and lung activity for a period long enough to produce beneficial changes in the body."[4]

DeCarlo certainly believes that aerobics helps to prevent heart attacks, and the Xerox Laboratory for Physical Fitness, for executives, has an outstanding aerobics program. Xerox has preliminary physical examinations, individually prescribed programs, and continuous testing and supervision. There are six activity stations in the Xerox program: warmup calisthenics, the treadmill, the body development machine, a bicycle exerciser, another turn at the body development machine, and then rope jumping. DeCarlo adds that "jogging is the most important factor in our program." Obviously millions of others believe in the importance of jogging and, fortunately, it does not require expensive equipment. Merely jogging, however, is not considered to be adequate for physical fitness.

Aerobics was initiated by Kenneth H. Cooper, M.D., a former Air Force medical officer. Dr. Cooper continues to research and write on aerobics. He is the head of the Aerobics Center in Dallas and is engaged in long-term research to establish for certain whether aerobics does, as he believes, reduce the incidence or damage of heart attacks and serve as an aid in the recovery from such attacks.

Total Isokinetics

Another physical fitness approach which has its supporters is "Total Isokinetics." Professor S. Brown of the University of Arkansas provides this explanation:

> Total isokinetics combines static and dynamic exercise. The muscle group being exercised is subjected to a maximal 10-second contraction, followed immediately by a 12-second movement

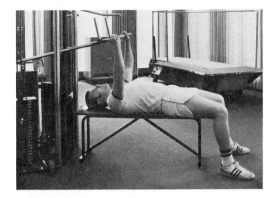
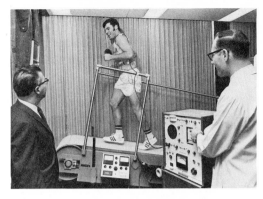

Isotonic and Aerobics stations at the superbly outfitted Xerox Laboratory for Physical Fitness.

through a full range of motion against predetermined resistance. . . . The static contraction reduces blood flow into the exercised area. The oxygen in the tissue is exhausted, and the muscle fiber quickly tires. Then, without relaxing, the fatigued muscle is worked through its range of motion in the manner described. The end result is to lower significantly the number of repetitions needed in a specific exercise to produce the desired training effect. This, in turn, reduces the total amount of time needed for a daily workout.[5]

Total isokinetics has been established at several high schools and colleges. It is also used by several professional athletic teams. Professor Brown writes that "for the NHL's Philadelphia Flyers it is the sole means of conditioning."

As compared to aerobics, total isokinetics requires less time and no heavy stationary equipment. Research on the effectiveness of total isokinetics is being done at the University of Arkansas and at Texas State Women's College.

Other Approaches

Aerobics is the most popular system and most fitness clubs use aerobics or their own modified version of it. Some develop their own programs. However, special systems are not the complete answer because not all companies provide such a program, not everyone can afford them, nor can

Women usually choose physical fitness for slenderizing. Here are slimming classes without equipment at Pratt and Whitney Aircraft Club.

they attract all those who need physical fitness. Dr. Kishore Ambe believes that "firms ought not to . . . create the impression that employees should do all their fitness work in a gym or other company facility.[6]

According to C. Carson Conrad, Executive Director of the President's Council on Physical Fitness and Sports, jogging is very important and should be included in every program. "As a conditioner," he notes in the Council's brochure, "there is no activity better for the cardiovascular system." Jogging, swimming, bicycling, hiking, and other activities in or about the home can do wonders and should be employed more regularly in individual fitness programs.

Motivation for Fitness Programs

It is difficult to motivate a high percentage of employees to enroll in fitness programs, and keeping them in such programs is even more of a problem. There are a number of recognized motivational techniques which can be employed in starting a fitness program or in interesting new or additional employees. These include: (1) factual and scientific presentation of the benefits of physical fitness, (2) the use of inspirational films and filmstrips, (3) visits and addresses by well-known sports figures, (4) provision of a company facility, and (5) the enthusiastic support of top management.

Malcom S. Forbes, President of Forbes, Inc. and president and Editor-in-chief of *Forbes Magazine* is a sporting enthusiast of world reknown. He is responsible for the Fitness Center on the fourth floor of the company headquarters and for the additional athletic facilities on the roof of the building on New York's Fifth Avenue. His enthusiasm and personal interest must contribute to the participation of the employees in the Forbes fitness program in which "approximately one-third of all New York based Forbes employees make use of the Fitness Center two or more times each week."[7] That participation rate would be the envy of most companies, but not satisfied with that, Forbes is supporting a research study on how to increase participation.

One Research Project

There is at least as much difficulty in keeping persons in fitness programs as there is of getting them in at the start. In an effort to find out how to keep them in, Robert S. Wanzel,

Ph.D., a contributor to this text, and a colleague of his at Laurentian University, Richard R. Danielson, Ph.D., decided to investigate why participants withdrew.[8]

Wanzel and Danielson found that twelve basically different reasons were given for program withdrawals. The leading reasons, running almost neck and far out ahead of the pack, were: (1) the distance of the facility from the home or place of work, and (2) the interference of the workout period with their daily schedules. Further back were listed: (3) workout crowded, (4) medical problems, and (5) the workout program itself. Other reasons given by fewer respondents included, in no particular order: the instructors, the locker rooms, the equipment, injury, and the attitudes of either supervisors, fellow workers, or of the family.

Most dropouts came within the first six months and included almost all (93.2%) of those who had not gained their stated objectives. During the same period 65% of the respondents who maintained that they had achieved their objectives also withdrew from the program.

Wanzel and Danielson found that the objectives most often stated at the beginning of the program were: general conditioning (70%), improved heart and lungs (52.8%), and weight loss (46.5%). Some respondents gave more than one objective. The first two objectives obviously were too general.

An interesting finding was that participation was higher in the fall and winter months. Another was that many persons withdrew subsequent to their summer vacations. Significantly also, though the total number of complaints were not large, considerably more females than males withdrew because of the instructor.

Research Conclusions

In consideration of the reasons given for withdrawal from the program, Wanzel and Danielson made several suggestions aimed at lowering the rate of withdrawal:

1. Incorporate the fitness program within working hours. This would eliminate the two reasons most often given for withdrawal.
2. Try a "buddy" system. This might add personal and psychological support.
3. Encourage participants to "develop a realistic objective or step that can be attained within a two-week period." This should be followed, objective by objective, for at least the first six months. "An objective like general conditioning should be the end result, a by-product, of all the steps attained along the way."
4. Instructors should make a particular effort to encourage and help female participants.
5. After vacations of participants, the fitness director and/or the "buddy" should approach the employee and encourage renewal of fitness participation.

Other Motivators

A number of other motivational ideas to encourage continuance in physical fitness programs have been suggested from various other sources. We list several which may be helpful in some circumstances.

1. Too often physical fitness programs are imposed or there is a certain amount of "arm-twisting." It would help to get more employee involvement in planning and operation. Enthusiasts, electing their own officers and working together, would probably aid in maintaining interest and participation. Form a fitness club!

2. Provide both indoor and outdoor facilities.

3. Vary the fitness program with activities in addition to, or sometimes in place of, the specified exercises or exercise equipment, swimming, for instance.

4. Arrange for competition between work divisions or with other recreation clubs. Competition might include percentage of sustained participation over specified periods. It might also include specific and objectively measured objective achievements.

5. Make provision for public recognition of those reaching certain specified and measurable goals. This could be done by the presentation of certificates or other inexpensive tokens, preferably by someone high in the executive ranks of the company. Also include mention of award winners in association or company newsletters.

6. Persuade top management personnel to exercise with the troops. This would be a tremendous morale booster.

7. Permit both sexes to work out together. Since objectives are more or less individual, this does not create unfair competition. It would permit spouse participation with mutual reinforcement. It also might make the scenery better and the workout periods more interesting. There may be reason for separate workout periods part of the time.

Whatever the areas of disagreement about physical fitness programs, there is unanimous agreement that physical fitness is a goal worth achieving and maintaining. The larger the number of participants in a particular program, of course, the greater the challenge of scheduling— infinitely more pleasing than the problem of increasing participation. And scheduling is, not by chance, our next topic.

Scheduling

It may appear that this book, along with *Recreation Management,* tend to give a disproportionate amount of attention to the larger recreation programs. Those programs do have the greatest opportunity for variety and for new activities. Also they have the largest professional staffs which write and publish. Therefore, more information is available about them.

In this section we will reverse the order and begin the topic of scheduling with the criteria for smaller clubs having few or no facilities of their own. Nonetheless, many of the suggestions made for them also have validity for associations or larger size also.*

During the last few years as the scope of activities has broadened in industrial recreation, the area of scheduling has become a whole new ball game. No longer is the only consideration whether Team A plays Team B on Tuesday or Wednesday night. The more prominent influences today include: facilities, community programs, work schedules, and priorities of employees.

1. Facilities: Programs which must rely on the facilities of schools and municipal and private recreation must vary their own schedules accordingly. The industrial recreation director somehow has to accomodate the times unused by the facility owners with the priorities and work schedules of the club membership, not an easy job.

2. Community Programs: "The director does not want to compete against community programs by scheduling like events at the same time." Some club members live within the community

*This procedure is possible because of the excellent manuscript provided by George D. Mullen, CIRA, Recreation Director for Frigidaire Employees Recreation Association, Dayton, Ohio. Most of the ideas referring to small program scheduling are his as are all quotations not otherwise identified.

and wish to participate in both programs. "Intelligent scheduling will permit both and, not only provide the employee the opportunity to participate in recreation activities of their choice, but to broaden their social life through these additional exposures."

3. The Work Schedule: The Company's work schedule has to be a factor. "The frequency of overtime and its duration are prime factors to be considered." Nevertheless, it is feasible to schedule events so that hourly and salaried employees can participate in the same activities. Second and third shift employees' activities deserve, but often do not receive, the same considerations in scheduling as do the first shift.

4. Employees' Priorities: Increasingly a number of employees are participating in two or more activities. Activities often run concurrently or overlap with each other, as the seasons for softball, bowling, and golf. This may be countered by: a change in the night a league will play, reducing the length of the season, or starting league play a week earlier or a week later.

Most members will have a priority of preferences for activities. The tendency is for members to give up the activity in which they are least skilled and thus avoid enriching interests and variety of activities.

"Every director will find variances in these and other factors influenced by the company's policies, community involvement, areas of the country in which they are located, et cetera. Scheduling is effective only when the maximum number participating is the maximum number available to participate."

Other Scheduling Matters

There has been a great deal of discussion concerning the scheduling of physical fitness programs. The President's Council recommended in 1973 that the fitness program should be on company time, both for motivation and supervision. Robert Wanzel's doctoral research and subsequent studies lead to the same conclusion. This is not at all a radical idea. As Gerry Glassford, Ph.D., of the University of Alberta, Edmonton, Canada, observed: "In some parts of the world exercise breaks are as much an institution as coffee breaks in the United States and Canada. Industrial studies have proven that overall productivity increases when recreational or physical activity is introduced at critical periods during the working day."[9]

At the very least, the fitness gym should be open during many hours. In April 1972, Martha Daniell, CIRA, then Director of Recreation for Nationwide Insurance Company at Columbus, Ohio, reported a 24-hour arrangement:

> The facility gets its heaviest use at lunchtime breaks, although it is always open: some Nationwiders like to exercise before work, others prefer to wait until the working day has ended, and some data processing personnel on the swing shift use the facility in the middle of the night.[10]

Fitness facilities are not the only scene of scheduling problems. Even the largest and best equipped of other recreation facilities have their share of scheduling problems. The Kodak Park Activity Association program found that interest in squash grew to the extent that the reservation of court time grew expensive and inefficient. As a result the KPAA devised a sign-up system under which participants fill out a monthly court scheduling request for court times. Selections are made at random and all participants are accommodated as fairly as possible.

When the large increase in sports participation by women put pressure on already tight gym schedules, other accommodations had to be made. A new rotational system scheduling volleyball teams at varying times for different weeks was one solution. In the basketball league a "bye" system was introduced whereby teams rested once in every few weeks. In softball, it became necessary to schedule two games per evening on available unlighted fields. Teams rotated early and late times. The regular season schedule was shortened and a week-end tournament was added.[11]

Scheduling problems are eased if calendars are developed well ahead of time. An annual calendar is desirable, giving at least the highlights. The greater the detail, the better. Such calendars in one form or another are found for many programs. Sometimes they give not only the date and time but also the place. This is a wise provision to help eliminate conflict of facility use.

The Burns Harbor Activities Association of Chesterton, Indiana, publishes a yearly calendar of Special Events as well as the meetings of subclubs and tournaments. The Flick-Reedy Company Newsletter carries a list of events for the coming months and a calendar with spaces for the employees to mark their plans.

In all schedule planning, widespread involvement is recommended. A tentative calendar circulated first to the governing board and then to subclub chairpersons will almost always reveal some errors and unnoticed conflicts. These can then be resolved before the final schedule is printed and circulated.

The one aspect of recreation clubs which is not usually affected by scheduling problems is club services. That is why services have been left to the last section of this chapter.

Services

Lastly we come to the employee services of industrial recreation. Service may not be as exciting as the sports and other activities, but it is the bread, meat, and potatoes of really successful employee association programs. It can be exciting for those recreation directors who welcome a challenge and who have a large streak of the milk of human kindness. The growing recognition of the importance of employee services marks the beginning of the end for the oldtime ex-athlete recreation director who knows nothing and could care less about anything other than sports.

During the last few years more and more of the services formerly performed by the company personnel offices have moved to the offices of the employees association. The new-model, professional employees activities and services director welcomes the transition. It represents an opportunity to demonstrate the value of the employees association to the employees themselves and to the Company management. It provides new appreciation and the likelihood of additional financial support from both.

Everyone well acquainted with the field knows that more employees participate in the association's services (if extensive and well performed) than in all the other aspects put together. It is number one! That at least some professionals are very much aware of this fact is obvious in an article written by three of them.

> Anything that is an additional service to the employee is better for recreation's image to management and employees.
>
> The clever, aggressive recreation director should take full advantage of his new status. The less aggressive director, who feels all his employees need are sports, will find his programs possible in jeopardy—the very survival of his recreation department threatened. Sports programs alone simply do not involve enough people and cannot generate enough money to help your programs survive financially especially in these [1975] times.

. . . Recreational professionals must take advantage of this change with the times. For the ambitious, innovative recreation/employee services manager, these are terrific times to expand and professionalize.[12]

The challenge of the service area is the greater because it is an aspect of the recreation program for which the manager must take the major, almost the sole responsibility. It requires continuous supervision, evaluation, and a year-to-year continuity that almost rules out operation by the elected officers of the association.

This should not indicate a lack of need for membership involvement; quite the opposite! The assistance of committees and volunteers to help in carrying out the service program should be encouraged because the program cannot be as effective otherwise. There should be an overall services advisory committee as well as subcommittees for particular tasks.

Services or Revenue

Here, as elsewhere, we run into some problem of duplication. Two activities discussed in chapter 5 as sources of revenue may also be important services to the employees. These are travel services and the association store. In some instances these exist chiefly as providers of revenue for the association, though, even then, they must also be of real service. Otherwise they are not likely to flourish. On the other hand, both travel services and merchandising are sometimes offered almost as pure services with the financial aim of only breaking even. Actually, it would seem that both profit and service are reasonable objectives.

The service aspect is emphasized in the employee store at Nationwide Insurance Company in Columbus, Ohio. Significantly there, as related by Bobbie Hildenbrand, "unlike most merchandise resale programs, which grow out of established programs, the effort at founding the employee store provided the impetus for the creation of the recreation association."[13] In Hildenbrand's view, "merchandise resale programs give employees a bonus, as surely as if the payroll department had slipped a few extra dollars into every paycheck." The Nationwide store, of which Martha Daniell, subsequently a president of NIRA, was one of the early supporters, was founded in 1950, with the cooperation and strong support of the Nationwide management.

Because travel services and association stores were discussed earlier, we shall confine ourselves here to other services, particularly of the kind which may be handled through a common service center.

The Service Center

For the most effective and beneficial services program, there ought to be a special Service Center. The multitude of services can not be well handled as a sideline of the secretary to the recreation manager. This, of course, depends upon the number of services offered, but if they are as great as is desirable for a large organization, the secretary will have too much else to do. Wherever located, the Service Center must definitely be under the supervision of the director of recreation, and his personal participation is necessary in some aspects, especially at their beginning.

There are several services especially appropriate for the Service Center. Among these are:

1. discount tickets and reservations
2. information services
3. loan services
4. sales
5. reservation of facilities
6. photo services

11. classified ads
12. hired hand register
13. drug services
14. auto services.

7. rental facilities
8. licenses
9. utility bills
10. floral service

Each of these require at least brief discussion.

1. *Discount services.* This involves discounts for tickets to amusement centers, special events, and movies or theatres. In some instances this will also involve reservations. Also there are many businesses which, though arrangement made by the program director, will offer all association members discount prices on merchandise. Some of these are local merchants and some are national distributors. Of the latter, most are NIRA associate members, of which more will be said later.

The Service Center is not itself a store (though a number of associations do have stores) and involves stocking little or no merchandise. It will have catalogs or information about merchandise which is available at discount rates to club members. Such items range from sports equipment and general merchandise to automobiles and, at least in one instance, land and home purchases. There is great purchasing power in large numbers.

The power of numbers is demonstrated by the story of "The Great Triumvirate." It relates the success of three large medical centers in New York City which, through the cooperation of three enthusiastic women, were able to get marvelous ticket services for the forty thousand students, nurses, staff, doctors, and technicians represented. They dealt in hundreds of thousands of tickets: some free, some at substantial discounts, some reserved in blocks of seats for whole departments.[14]

The immense leverage of large numbers is available not only to the very large recreation clubs, but to two or three who can cooperate as "the great triumvirate," and to employee associations of all sizes that are members of a local industrial recreation council.

In all circumstances in which the recreation club acts virtually as the middleman, caution should be exercised to see that what is offered really is at a discount and of good quality. It is not a good policy for the club to reject all responsibility and to operate on the principle of "caveat emptor" (let the buyer beware). The director and the club will get part of the blame for shoddy merchandise. John Ruskin said in the 19th Century: "There is nothing in this whole wide world which some man cannot make a little poorer and a little cheaper; and he who buys by price alone, is that man's lawful prey." The wise recreation director will spend time and effort checking out the "good deals" recommended to members. That is a large part of the service performed.

2. *The Information (Library) Service.* What has been said about special discount arrangements is not necessarily true when all that the Association is what might be called "library" services. If the information provided is clearly labeled as *only information and not a recommendation,* there should not be a problem. The Jet Propulsion Laboratory Employees Recreation Club (ERC) offers a modified approach which may work very well, but which does carry some slight danger. The following passage is from the ERC brochure.

ERC "Buyers Guide"
As part of its program of purchasing assistance, the ERC office carries Consumer Reports magazines, wholesale catalogues, magazines from Fedco, Homebuyer, Lens, and Los Angeles Magazine in addition to a list of firms either offering discounts or simply "good value for money received" as reported by fellow employees, always on the basis of "caveat emptor" (let the buyer beware) for the protection of all concerned. Many employees help the Executive Manager of the

ERC to keep this list up-to-date by suggesting where to purchase at the best price as well as organizations that should be deleted.

A list of car dealers who offer new automobiles at fleet discounts is also kept at the ERC office, where there are also copies of Kelley Blue Book for used cars and new car invoice prices. These cannot be checked out of the office.

Two particularly commendable features of the ERC "Buyers Guide" are: (1) the quality of such library services as *Consumers Report,* the *Kelley Blue Book,* and new car invoice prices; and (2) the involvement of the many employees in providing information. Nonetheless, we are inclined to the belief that the safest and wisest course is for the recreation director or a staff member to check out each recommendation.

Other library-type information sources that might be collected and made available by the Service Center include:

a. Brochures and booklets about vacation spots and crafts, how-to-do-it manuals for such things as auto mechanics and home electrical work
b. Maps and schedules: city maps, county maps, road maps, camping atlases, local bus routes and schedules
c. Local residential and business directories, and also directories—probably hand-compiles by necessity—of local recreational facilities, cultural centers, educational institutions, and all kinds of services available, including a hall-rental directory
d. Catalogs of nearby colleges, along with their brochures about current offerings, adult education courses, extension courses, and fees
e. Information, for parents and employees themselves, of sources of financial aid for higher education: special financial programs, scholarships, and loans, in addition to information about terms, eligibility, and where and when to apply
f. Social Security and Veterans Administration booklets

3. *Loan services.* This includes the checking out of equipment, especially for lunch-hour recreation, such as dart boards, checker and chess sets, playing cards, ping pong equipment, and perhaps, some sports equipment as footballs, softballs, and gloves.

Another type of loan service would be 8-mm movie projectors, movie editors, movie cameras, tape recorders, small projection screens, and 35-mm slide projectors for home use of for special presentations within or without the club.

A loan service that would be greatly appreciated includes convalescent items: crutches, walkers, et cetera. Many persons buy such items, use them for a short time, and store them or throw them away. Many would be willing to donate them to the recreation club. One club loans tools to its members. In fact, all the items mentioned are on loan by one or more recreation associations

4. *Sales.* A number of small convenience items might be kept for sale: greeting cards, playing cards, women's hosiery, golf balls, tennis balls, athletic socks, and loyalty items. The latter are small items either manufactured by the company or specially purchased with the insignia of the company or the recreation association.

5. *Reservation of facilities.* This includes the reservation of club meeting rooms as well as athletic areas such as racquet courts and tennis courts.

6. *Photo services.* If there is not an association store, the Service Center could handle, both as a convenience to its members and for association revenue, film, flash bulbs, and film finishing services.

7. *Rental of facilities*. The Association might rent its halls and banquet rooms to members at low cost for special occasions such as weddings, receptions, anniversary parties, and annual family meetings.

8. *Licenses.*The Association can render a good service to its sports enthusiasts by handling hunting and fishing licenses. In many states, it may also offer the convenience of handling automobile and boat licenses.

9. *Utility bills*. The acceptance and handling of utility bills would also be a much appreciated service. Aside from the convenience, it might save each member five or six dollars a year for postage and stationery.

10. *Floral services*. A list of florists with discount arrangements would be a convenience. It would be even better if the Center could include florists catalogs, and ordering and delivery services.

11. *Classified ads*. This involves registration with the Center which arranges for publication in the Company and/or Association Newsletters and the posting on bulletin boards of articles for sale or exchange. The bulletin boards need to be well managed and arrangements made concerning what is to be posted, the size of notices, and so forth. Also, each advertisement must be dated, and someone must be given the responsibility for removing outdated notices and keeping bulletin boards orderly. Only if they are neat, uncrowded, and current, will bulletins boards attract much attention and be of real communication service.

12. *A Hired hand registry*. This is similar to the classified ads service, but requires closer supervision and more work on the part of the director or staff. Essentially it is an annually updated list of persons who have skills available *on a part-time basis to members of the association.* Created by Martha Byers, CIRA and member of the NIRA Board (1978), the hired hand register is very successful at Owens-Illinois in Toledo, Ohio, where Byers is Employee Services Director. Originally, the lists of persons and their skills were created for the retirees of the Golden Emblem Club; but the demand became so great that the register now includes current employees and members of their families, and even neighbors who are recommended by OI employees. The services offered, however, are available only to OI employees and retirees. Among services offered are: plumbing and electrical work, tailoring, auto repair, house painting, typing, babysitting, invalid care, and occupying or maintaining a home during the owner's vacation. It is a wonderful service for every employee and retiree but "especially for older people who are alone and can't do everything for themselves."[15]

13. *Druggist services*. Such services are particularly important when, as is often the case today, both parents work. The Center can arrange for a nearby and dependable druggist to deliver prescriptions to the Center. Even though many drugstores have delivery service for homes, that does not help when parents are not at home to accept delivery. Moreover, prescriptions are often for the employees themselves and they would be delighted to pick them up at the Center during the day or when leaving. If it became a volume operation, discounts could probably be negotiated.

14. *Auto services*. Very frequently when both parents work or there is only one car, it is quite a problem to get the car(s) repaired. It is often hard to manage without the car for a single day, and it is often hard to get to a garage at times when the employee is not at work. Even if that can be managed, transportation is needed to work or to home and back to the garage. An answer would be for the recreation association center to arrange with a nearby repair service to pick up the employee's car at the work site, repair it, and return it in time for the employee to drive home.

A "car repair" parking place could be reserved and keys left at the Service Center. A repair service that valued the large amount of business generated by such an arrangement would be willing to give extra benefits such as pick-up and delivery, and perhaps also make a price allowance. Moreover, the repair service would try to do good work because of the large amount of repeat work possible. The same service company could be called upon for help in starting cars and for flat tires on the company lot.

Depending upon the number of employees and the number of services offered, the Service Center may become a large operation. It would cost money, but he idea should not be taken lightly. For some services, or all, if necessary, a small fee might be charged. Good management might allow it to break even. The chief criterion should be the excellence of service to members. A complete and well-operated service center should be enough by itself to induce a huge percent of the employees to participate in the association.

If such a center is inaugurated, the recreation director must see to it that the operation is efficiently organized and operated, that there is good record-keeping, and that the financial affairs are controlled and audited. Employees of the Center should be carefully chosen. They should have integrity, a cheeful manner, courtesy, initiative, and, desirably, a good-starting knowledge of the community. Their special abilities should include getting along well with people, working well under pressure, and being able to accept responsibility.

Instructional Services

A recreation association should provide a number of instructional services. Among these are instruction in crafts and creative hobbies. Such instruction may be provided by hobby clubs, once formed, but often instruction must precede the formation of the hobby club. Also many persons will not join an established hobby club until they already know something about the hobby. They do not wish to begin as a complete novice when, at least they fancy, all the other members are quite skilled.

The development of creative craft skills is an avenue to self-fulfillment. It is also preparation for retirement. Martha Daniell, CIRA, in a manuscript prepared for the book, particularly emphasized the role of retirement preparation.

> A well-rounded craft program can provide preparation for retirement training for many employees who up to this time have never applied their creative abilities to some worthwhile project. Perhaps they will learn ceramics, needle crafts, leathercraft, hatmaking and many other crafts that would even become a vocation or partial-income possibility after retirement. . . . Some retired employees have established small businesses of their own with skill acquired through the crafts, such as quilting, rug making, doll repair, woodworking, clock repair and finishing, etc.

The economics of the situation obviously recommend group instruction. This can be arranged either for free or at a charge sufficiently low to attract participants yet high enough to pay the instructor's fee. Sources for instructors for the crafts are the YMCA, the City Recreation department, craft and hobby stores. Also remember the high schools, many of which have good well-rounded vocational art programs. Museums and colleges are also a good source of instruction in many arts.

In some instances the most practical approach will be an adult education course provided by a local school or college but taught, if you wish, on your own premises.

A number of the better recreational programs do offer much in instruction, and we copied the following list from the 1975 Annual Report of the Kodak Park Activities Association.

Hobby class	Number participating	Hobby class	Number participating
Art	55	Boating Safety (jr.)	21
Bridge	150	Ceramics	110
Charm (Jr.)	75	Dancing	150
Dressmaking	30	Fly-casting	100
Fly-tying	30	Golf	278
Horseback riding	45	Hunter safety	168
Karate	50	Needlecraft	30
Scuba Diving	20	Sewing	60
Tennis	20		

Other instructional areas which have not been mentioned thus far but which we believe especially worth noting are: first aid training, life saving, auto clinics for women, and money

Cake decorating.

Auto mechanics for women.

Sculpturing.

Painting.

Some of the cultural classes offered by the Pratt and Whitney Aircraft Club.

management (budgeting, buying, personal banking, savings plans). Closely related is wise investing. Investment clinics may go into many areas, each taught, perhaps, by a different instructor. Investment areas include: stocks and bonds, land, real estate, antiques, old car restoring, and many others. In many instances investment clinics have been followed by the organization of investment clubs. Mutual investment clubs are a lot of fun, very instructive, and, often, quite profitable. Above all, they are an excellent way for beginners to get experience with minimal risk.

Special situations also call for special kinds of instruction. If in the Company there are very many non-English (or broken-English) speaking employees and/or spouses, a two-way language program might be undertaken. On program should be English as a Foreign Language. The other could be instruction for the English-speaking employees in the prevalent foreign language. Such courses probably should be only introductory or of the conversational type. Participants may follow up with more advanced courses offered by international institutes or by colleges.

There are many other avenues to helpful instruction. A suggestion made in the NIRA "Key Notes" by Mel Byers, CIRA, [V.2, No. 3 July 1978] was to start a University Club. Allying with

Archery.

Horseback riding.

Karate.

Rock climbing.

Some of the many physical activities and sports classes offered at P & WA Club.

a nearby university, you might arrange for a series of luncheon or dinner lectures pertaining to a wide range of subject such as travel, music and art appreciation, book reviews, and antique collecting. The list is endless.

Miscellaneous Services

There are a few other services that do not seem to fit in the other categories. They are rather widespread and common, but should be mentioned.

A Credit Union. Unless already supplied through the Company, a member credit union is an excellent service for an employee recreation association. It may start off slowly but is bound to grow in participation and usefulness.

A Blood Bank. By special arrangement with the local Red Cross Blood Bank, your organization members may make on-the-spot blood donations. The members of an association which participate in such a plan will then be entitled to free blood for transfusions whenever and whereever needed. The cost of blood for transfusions is not usually included in most health insurance programs and can be quite expensive.

A Hospitality Program. Special welcome arrangements and assistance for the families of new employees, especially those new to the community, will be much appreciated. That is an excellent way of introducing them to the employee association. A second kind of hospitality service is the making of arrangements for flowers and remembrances for employees who have extensive illness or a death in the family, plus recognition of happier occasions as births and marriages. It is the thoughtful little touch that may be the longest remembered.

Share-the-Ride Programs. This needs to involve registration, and perhaps should have been included under the possible functions of a Service Center. In addition to notices spread like other ads, there must be some other provision. A large, area-keyed map, which could be used as a reference point for locating car pools and rides, would be helpful. If the company is very large copies should go to all work divisions for bulletin board posting.

Garden Plots. A number of companies and recreation clubs are providing garden plots for employees or members. The entire area is plowed and prepared for seeding. Then it is marked off into small plots, perhaps one-eighth or one-quarter acre, for assignments.

Services for Children

Many, if not most, of the more well-rounded recreation associations provide for sports instruction and athletic participation of members' children. Most often there are leagues and the parents provide the team coaching. Individual participation by children, as in club swimming pools, is also common. Many clubs assist with Scout Troops.

Less common child-related services deserve a special work. The Flick-Reedy Company has a children's Day Camp, a little different than it sounds. One day a week for nine weeks the day-camper goes to work the parent, visits with them on the job, and eats lunch with them. Game time follows, the older children looking after the younger ones. This is followed by swimming lessons, a rest period, and a free-swimming period. At the end of each 9-week day camp, President Flick hands each participant a graduation diploma.

Flick-Reedy is also noteworthy for special programs for handicapped children. These children, according to their abilities, are assisted in muscular growth and coordination, and are taught swimming in the company pool. Except for one administrator of the program, the large amount of TLC (tender loving care) is supplied by volunteers.

We are led here to an important question posed in the April 1975 issue of *Recreation Management,* "Should Industry Sponsor Day Care Centers?"[16]

It seems from here that the larger recreation associations are missing a very good opportunity if they do not offer a children's day-care service. More and more mothers are joining the work force, and day-care centers are mushrooming. Many businesses and universities are finding it to their advantage to sponsor nonprofit day-care centers. Such centers are too expensive to be offered as a free service; but they can be operated without loss to the association and with lower charges to the parent members than charges of private centers.

Child care costs for working parents are now federal income tax deductible, and as a service, wow! Advantages to the employees are:

1. No need to leave home early to drop the child off at a day-care center.
2. Fewer worries. The parent is at hand and can look to the welfare of the ill or injured child without missing the remainder of the day's work.
3. No need to worry about getting started home at the earliest possible moment to pick up the child. Arrangements can even be made for some overtime work.
4. The parent's morale is higher from the knowledge that the child is more secure and happier so nearby. It may be feasible for the parent to look in on the child during the lunch time.
5. Benefits to the Company are substantial. Absenteeism will decrease, and, with the better emotional condition of the parent, productivity should rise. Certainly morale will be better.
6. Benefits to the recreation association are the grateful appreciation of parent members and the likelihood of better participation in other association activities.

There are some cautions. If a day-care center is to be considered, attention must be paid to all federal and state provisions for zoning, health, and those regulations especially aimed at day-care centers. The zoning laws will more likely be local, as will fire regulations. Above all, in the planning and in the operation, a professional director must be employed. Depending upon the size of the operation, professional technicians may need to be hired, but volunteers will come forward and can be extremely helpful under professional direction.

Do not be deterred by the need for at least one professional and the considerable expenses of establishment of the center. The day-care center definitely can be a self-supporting service.

Service to Retirees

There are at least five steps to be taken in preparation for the enjoyment of retirement. To what extent the recreation association need to be or will be involved depends upon the extent to which the company is already making provision through its personnel department. The five steps are:

1. Learning to enjoy an activity that can continue after retirement. Creative hobbies are among the best. With its hobby clubs and craft instruction, the employee association is best able to handle this one.

2. Learning a skill which may be productive both in interest and financial revenue after retirement. Social Security and company retirement plans together are often not adequate to maintain the retiree at the level to which accustomed. The fastest growing poverty class in America is the old persons.

3. Beginning a planned savings and investment program several years before retirement. The association can contribute through financial education and investment clinics.

4. Receiving personal counseling near and at the time of retirement. What are the provisions of Social Security? Medicare? The Company pension plan? Health insurance? What about relocation? Many questions suddenly become vitally important.

Personal retirement counseling is most often provided by the Company's personnel section, and should be because the Company has the greatest reservoir of expertise in financial planning, legal matters, and government and Company provisions for health care and income for retirees.

There are several outstanding operations of this sort. Pratt and Whitney Aircraft begins preretirement counseling in voluntary group sessions for employees of 62 years of age. The company follows with phase two counseling at the age of 64 and concludes with personal counseling at age 65, the time of retirement. Personnel at the given ages are personally invited to participate and given time off with pay during working hours to do so. Spouses are also invited.[17]

There are undoubtedly other good retirement programs, but just as surely, there are some companies which do not take their responsibilities to their retiring employees all that seriously. If the company does not do the job of personal counseling, the employee association ought to do what it can.

5. The fifth step, and one of the most important, is involvement and activity after retirement. That is where the recreation associations can and very often do shine. Tragically, the retirement to which so many people look forward with such anticipation turns out to be traumatic in the extreme for some persons, most often males. Studies have shown that one out of every twelve persons of retirement age lack family connections. Many do not even have close friends. This helps to explain the heavy incidence of suicide by men recently retired.

Retirees and the Association

The industrial recreation association can keep the retiree from feeling completely alone and abandoned. Among the things which many associations do are: Give the retiree life membership in the Association. Invite the participation of retirees in regular club activities. Have annual special welcome back days for retirees. Develop a special club for retirees and assist it to promote activities and hobbies. Include a retiree on the Association governing board so that there is continual input from that segment.

Retirees can be a great asset to the recreation association. Many of them remain with abundant energy and enthusiasm, and they have the time for the volunteer activities which are helpful to the association. In every way possible, and it is not always easy, keep the retiree involved and active in the Association. Remember, all you have to do is live long enough to be one too.

A happy meeting of retirees at Pratt and Whitney.

A good, ongoing program for retirees is one of the most subtle forms of good public relations. It also is an excellent recruiting device. People like to work where a company and its employee association do care about people as individual human beings.

Caring about people also is important in respect to the next chapter, which is about program administration.

Notes

1. "New Developments in Maintaining Health and Fitness." *R.M.* April 1972, pp. 6–9, 28.
2. "Monthly Vital Statistics Report." U.S. Department of Health, Education, and Welfare, National Center for Health Statistics. Vol. 26, No. 1, 1 April 1977.
3. "Monthly Vital Statistics." Vol. 25, No. 2, 30 April 1976, and Vol. 26, No. 5, 10 August 1977.
4. "Saving Lives by Stress." *R.M.* April 1972, pp. 10–12. DeCarlo, Recreation Manager for Xerox and Past President of NIRA, was highly instrumental in the development of this text.
5. "Total Isokinetics." *R.M.* July 1976, pp. 20–22.
6. "New Developments." *R.M.* April 1972.
7. Kenneth Tillman and Curtis D. Cleland. "Up on the Roof." *R.M.* July 1976, pp. 23–25.

8. "Improve Aherence to Your Fitness Program." A Paper presented at the NIRA convention at Orlando, Florida, September 1977. The address was subsequently serialized in *R.M.* in the issues of July, August, and September 1977.

9. "The St. John's Edmonton Report." November 1974.

10. *R.M.* April 1972, pp. 16–18.

11. Kirk T. Compton, CIRA. "Organization Profile: Kodak Park Activities Association." *R.M.* September 1977, pp. 12–15.

12. Mark Armstrong, CIRA, Dick Brown, CIRA, and Jerre Yoder. "Employees Services—What Your Employees Want, Is What They Should Get." *R.M.* August 1975, pp. 8–9.

13. "Helping Employees Stretch Their Paychecks." *R.M.* April 1978, pp. 20–22.

14. Rose Migliore, Inez Greenstadt, and Patricia Byrne. *R.M.* March 1977, pp. 18–20.

15. "Hired Hand Register." *R.M.* March 1978, pp. 14–15. See also "Small Job Clinic." *R.M.* October 1973, pp. 20–21.

16. Henry Diehl, Coordinator of Recreation Curriculum, Triton College. pp. 12–13.

17. *The Hartford Times.* 9 June 1974. Also Pratt & Whitney retirement folder and brochures, courtesy of Von Conterno, CIRA, Manager Pratt & Whitney Aircraft Club, and R.D. Baikal, Personnel Counselor, Pratt and Whitney.

8
The Director

TOP MANAGEMENT SPEAKS

Industrial recreation is an extremely important activity at Motorola. I have given it a great deal of personal attention and have participated in many of the programs. My participation has been good for me, as I am convinced it has been good for my associates.

The basic objective of industrial recreation is to recognize man's needs as a social entity. This is a sound objective. Employee recreation has given opportunity for personal expression, individuality and recognition to the men and women in industry. This is an essential part of belonging to an industrial organization.

Through recreation, employees become better acquainted. We all find that we have many interests in common with one another. We gain better understanding of each other. We break down and overcome the persistent barriers that God seems to have given us as obstacles to overcome. In the process, people have fun.

Aside from personal development, recreation is also good for the corporation. Being known as a company with a varied recreation program helps in recruiting. Further, employee recreation is of great value to families and to the community. We hope that our recreation program helps bring family, community, and company closer together.

New leaders are frequently discovered as the driving forces in company recreational

Robert W. Galvin

pursuits. Altogether, Motorola's recreational activities are a natural, mutually enjoyable extension of wholesome, on-the-job relationships. Our program has long been one of the strong links in the communications chain among all of us. I feel confident this will continue with ever greater vitality and enthusiasm, in the years ahead.

Robert W. Galvin
Chairman of the Board
Motorola Incorporated

It is impossible to describe accurately the position of the chief administrator of industrial recreation programs because the position is so varied in different companies. Some of the variation is suggested by the different titles used: director, supervisor, coordinator, executive secretary, manager, business manager, and other. (For simplicity we shall use "program director" or "director.") It is believed that the positions are more similar in terms of responsibilities than the titles might suggest. To some extent each administrator creates the position filled, and the person designated as "business manager" may by initiative and ability become the catalyst for program development and the hub about which the recreation program operates.

In too many instances the situation is further complicated by the assignment of other responsibilities such as those of personnel director or assistant, public relations manager, employment manager, editor of the company newsletter, or employee benefits coordinator. There are others.

In the larger companies and in smaller ones which fully realize the benefits of industrial recreation, the position of director will be a full-time job and there may be professional assistants as well as a technical staff. The professionally prepared industrial recreation director can expect in time to become the head of an enterprise which is large in numbers, complex in its organization, and varied in its operations. Additionally there may be extensive facilities, many employees, and an annual budget running into the hundreds of thousands of dollars. It is a responsible and challenging position, but it does have its difficulties.

Difficulties of the Position

The major difficulty is that the director of the recreation program is usually hired by and paid by the Company. The director is expected to be responsible to the Company and loyal to its interests, yet the director must also be responsible to the governing board of the recreation association and loyal to its interests. We find the crux of the problem described in a manuscript submitted by Jerre Yoder (Chief of General Dynamics Employee Services):

> The Recreation Administrator is caught in a vise, or becomes the rope in a constant tug-of-war between factions. He is in fact held responsible by Company Management, the Association Board of Directors, and Association Members for successful operations even though given inequitable and sometimes impossible parameters and circumstances within which to function. In essence, the Director is charged with almost total responsibility while being given very little authority.

Putting Out Brush Fires

Not all the tasks of the program director are important matters. A surprisingly large proportion of time must be spent in "putting out brush fires." These are the unanticipated developments, the small crises which demand the director's attention. Time must be spent in settling differences between members of the association or of the staff, in adjusting the inevitable personality clashes. Time must be used to deal with real or imagined grievances, to correct the inequities which sometimes rise from rules and procedures that are fair and workable in normal situations. Also there must be time for the correction of misinformation which threatens to imperil an activity or the entire association. These *important minutiae* consume valuable time, drain energy, and create frustration. They call for the patience of Job and the wisdom of Solomon, and they must be survived before the director is free to be designer, planner, creator, and executor.

Nevertheless, there are important responsibilities which almost all program directors share in common.

Responsibilities of the Director

Many of the responsibilities of the program director have been noted in previous chapters and will be given little or no further attention here. Among these responsibilities are: leadership in the development of goals and objectives, the creation of a workable organizational form and by-laws, program development, scheduling, and financing. There are others aplenty!

Leadership

Perhaps the greatest contribution of the program director will be to provide leadership. Leadership is the ability to strike a spark, fan it into a flame, and control its burning. Qualities of leadership include vision, imagination, ingenuity, creativity, initiative, and enthusiasm. On the whole, leadership is marked by the ability to get things done by others. The able leader is more catalyst than performer. Yet providing leadership is not the only role of the director. That position also requires other qualities related to administration, such as practicality, efficiency, and the capacity for details. Though we regard leadership as the most important role, the successful director must also be an able executor.

The Support of Top Management

A special, personal responsibility of the program director is to gain and keep the support of the company's top management. The toleration of top management is not enough; there must be active support if the industrial recreation program is to flourish. Such support is critical even for the incorporated and theoretically independent employee recreation association.

Being employed by the company and responsible to some management position in the company, the director is provided with a ready avenue for communication with top management. Also almost always Management is represented on the employee association governing board. Company representation may range anywhere from one ex officio, nonvoting adviser to the total membership of the board.

In respect to top management in general, the program director must keep lines of communication open and see to it that Management is aware of what is going on in the association. This includes past successes, current efforts, and ongoing plans. This should be a deliberate campaign employing personal contact, telephone calls, written reports, and the use of outside newspaper publicity and in-plant newsletters.

Whether many or few management officials are on the governing board of the association, the director must seize upon board meetings as an opportunity to impress. Before the meeting the director should research carefully the programs or policies to be recommended there. As a result, recommendations can be supported by hard data, facts rather than guesses or pure opinions. The director should be working with the association committees, and at the board meetings should be able to present proposals as committee proposals rather than as personal recommendations. It is unnecessary to add that the board meetings should be used to keep the board members, including management representatives, up-to-date on association matters and accomplishments. Beyond selling the association through communication, the director must sell Management by working with Management.

Working with Management

Ideally, the director should be able to work *with* Management as calmly and effectively as with rank and file members of the association. Working with Management means cooperating in the achievement of some Company goals. The director can support Management in many respects, particularly in such community efforts as blood-bank drives, community chest drives, and other public relation enterprises. The director can provide Management with an invaluable opportunity for communication.

Communication*

Communication is the main social thoroughfare from here to anywhere. Unlike a pedestrian thoroughfare, it cannot be traversed alone. There is no communication until two or more persons share information, attitudes, and emotions. There must be a transmitter and a receiver, both turned on. Though a radio station may broadcast all day, no communication occurs unless someone has a radio turned on and tuned in.

The radio is a useful simile, for we need to keep in mind that we also can be turned off or tuned out. We are most likely to be tuned out if we are threatening in manner or if the message is threatening. We also risk being tuned out if the message is dishonest and insincere, if it is very biased and consists of half-truths, or if it is exaggerated and overembellished. Stated in the positive, we are most apt to communicate when we are friendly and nonthreatening, honest, factual, and complete.

Communication involves emotions as much or more than logic. This is particularly important in industrial recreation where so much communication is between individuals or among small groups. Therefore, to the extent possible the director will try to know the receivers' backgrounds, preconceptions, and interests. Beyond that the director should strive to be perceptive, aware of current and changing attitudes and feelings of others. In communication it is important to seek feedback. We question, listen, and probe to discover whether what is coming out of the communication pipeline is what we put in. There is so much room for misunderstanding or misinterpretation. Only when the sender and receiver attach the same meaning to the words spoken or written does communication occur.

Finally, communication is more than the sum total of words and gestures. We communicate optimism by being optimistic, friendliness by being friendly, complacency by being complacent, and despair by despairing.

A Communications Network

The employee association can become a communications network if the director plans and works toward that end. In that case, the director and the director's office become something akin to a central switchboard. The director should maintain communication with top management, all supervisory levels, and the different worksites and operating divisions. Being at the same time in communication with the association governing board, the volunteer workers, the committees, the subclubs, and the members at large, the director not only passes information from one to the other, but stimulates each part of the network to seek communication with other parts.

All aside from the director's efforts, much unplanned communication occurs as workers from different levels and work areas get to know one another through association sponsored hobbies or

---

*The authors include here only a few bits of essentials of communication. Persons who plan a career in industrial recreation should have at least one complete academic course in the art and techniques of communication. The same is true of later topics in this chapter: personnel management and business administration. Our aim is to indicate a few essentials particularly applicable to industrial recreation and to stimulate interest in further study.

other recreational interest. For the most part, however, this important function of the employee association is not recognized. Program directors must take it upon themselves to help Management to discover that the employee association is a ready-made and highly-effective channel for communications of all kinds, moving in all directions.

Communication within the Association

Within the organization, one key to communication is involvement. That program is best advertised and promoted internally which has the interested planning and participation of the largest number of members. Each member does not have to participate, but each member needs to be represented. The different age groups, the different plant status levels, the different marital statuses, the different shifts, the different sexes—persons of all these categories should each be represented. This is both so that their ideas may get into the "mix" and so that they may receive feedback about matters which may be of special interest to them. If the employee activity organization has been organized with a broadbased pyramidal structure, and if it has been organized with communication as a deliberate objective, communication will be greatly facilitated. If communication has not been structured in deliberately, the director must improvise as possible.

It is especially critical that those in a leadership role within the association be kept well informed. This includes the members of the board of governors, the staff, committee chairpersons, and subclub leaders. The director should be especially diligent to keep in close communication with members of the board of governors. The formal board meetings are not nearly sufficient for adequate communication. These should be supplemented with personal telephone calls and informal notes. The importance of these informal contacts with other volunteer leaders differs only in degree. Being "in the know" makes all these persons feel better and more cooperative. It also makes them function more efficiently in the communications network.

Communication with members of the association staff is taken for granted. Yet it is surprising how often one staff member is ignorant of an assignment given to another staff member, even when that assignment may have implications for the work of the one left in ignorance. This is as true in business organizations as it is in recreation associations. This failure of the right hand to know what the left hand is doing is accidental rather than planned, but it will happen unless deliberate methods are taken to avoid it.

The importance of a well-informed staff can hardly be exaggerated. Failure to keep the staff well informed offends the members, leads to their embarrassment, creates unnecessary friction between the members, makes it difficult for them to function effectively, and even affects the director's credibility.

The best means of preventing this unintentional information-gap is frequent and regular staff meetings. These meetings also serve to increase the communication of staff members with each other. Staff meetings are opportunities to hear *from* the staff as well as to speak *to* the staff. The meetings are a way to keep all informed, to probe for feedback, and to pool knowledge and ideas. One hour a week (more or less, but certainly not two or three hours) should regularly be set aside for such staff meetings. Equally important is the "tone" of the meetings. An attitude of openness and receptiveness to the ideas of others creates the most fruitful climate for staff meetings.

Communication in Meetings

Of meetings, within the requirements of brevity, it is a challenge to write anything that is not trite or superficial and that will be especially applicable to the position of program director. Almost everything depends upon the size of the meeting, its composition, and its purpose. All are relative and exist on a continuum. Size may range from two persons to several hundreds. Other factors vary likewise.

Large Meetings

In employee recreation associations, there are few occasions for large meetings. Those few require careful planning both in regard to the program and the housekeeping arrangements. Program planning usually receives much attention but the opposite is true of housekeeping arrangements. Carelessness here can almost spoil a meeting even before it is begun.

Far too often people begin to arrive at a meeting to find the doors not yet unlocked and no one available with a key. Or the lights are not on, and those present do not know where to kind the right switches or fuses. Or the meeting hall is not yet sufficiently warmed, or cooled. Or the speakers' platform is not yet arranged. Or the public address system not set up and tested. The audience then may be irritated by not being able to hear, or by being assaulted by the squeal of microphone feedback.

Among the most common of omissions is failure to arrange for ushers. The speaker should not have to intersperse the message with announcements that there are still a few seats on the third row left. If these and other housekeeping arrangements are well handled, it is unlikely that the program director will receive much credit; but it is certain that the director will be blamed if they are poorly handled.

There are a number of other and less obvious responsibilities of the program director. The director will seldom conduct a large meeting—most often the function of the association president—but will always be very involved. It is the director's responsibility to train the meeting chairperson, if need be, in parliamentary procedures. The director also arranges for an association parliamentarian to assist and support the chairperson. It will most likely be the director also who will have to see that appropriate information is obtained for the introduction of an outside speaker.

The director must be prepared discreetly to rescue the inexperienced chairperson thrown momentarily off balance by a question, objection, or other unexpected development. Discreetly also, the director should serve as a moderating influence if a wide divergence of opinions develop.

Above all, the director must be well prepared personally to respond to requests for information or reports. Hardly anything is more impressive or persuasive than hard data, the exact facts and figures. Having that hard data available is crucial. Some can be memorized, some can be on note cards, and the remainder in printed material at hand. The director should be so familiar with that material that needed information can be located at a moment's notice. Being well prepared is not difficult; it merely requires careful planning and foresight. Some of the information needed will be made obvious by the program or agenda. Some will be suggested by personal conferences with the program chairperson. The remainder must be the result of deep thought by the director as to possible directions the program might take.

Large meetings are not the best method of communication, but they are important to the program director. If they are not carefully planned in respect to housekeeping details as well as to program content and actual presentation, they may be of very little usefulness.

Oral and Written Communication

Here we are concerned particularly with goal-directed communication. By "goal-directed," we mean communication which is undertaken to achieve a specific objective. Much of such communication will be directed to Management and is especially significant in that regard. Examples include requests for Management support for a new facility or a new staff assistant, Management support for a new program or a new program or a new policy departure, or special Management participation in some project.

Adroitness in goal-directed communication will contribute greatly to the success of the program director personally and of the employee association. The two basic avenues, oral and written, each have their advantages and disadvantages.

Oral communication is usually best for the introduction of new ideas or projects. Oral communication permits us to "ease" into a sensitive matter and then proceed or back off gracefully if Top Management appears unreceptive or unready. If the goal is important, we back off only to try a different approach later or to seize upon a more favorable time.

The oral approach also allows us to expand upon our ideas, clarify ambiguities, and respond to objections in the briefest period of time. In the event Management's response is enthusiastic, the oral approach allows us to seize upon the opportunity—to strike while the iron is hot—and derive the maximum benefit.

Conversely, a written proposal for a significant new departure often may carry a suggestion of bluntness or may be subject to misinterpretation. A rather innocent request is sometimes taken as a challenge or as a threat by a superior. An unfavorable response is also likely to be final and not permit of another try from a different approach later. The greatest weakness of the oral approach is that verbal understandings or agreements are too subject to misunderstanding and too dependent upon the frailty of human memory. Therefore almost all oral understandings or agreements should be put into writing.

Enthusiasm may wane and human memory is not very dependable. Having received a sought-after response from Management it might be tempting to say, "Put it in writing," but that is not practical. Sometimes Management will do just that on its own initiative. Lacking such a written record, the program director may yet get the matter into writing. The method used will depend upon many factors.

One direct approach is a simple memorandum of confirmation, prepared as soon as possible after the director returns from the meeting and while the matter is fresh in mind. One must be tactful, of course, but if Management does not immediately answer with a memo "correcting" the director's impressions, the director's memo remains permanently as the accepted version of what transpired.

An indirect and often more tactful method of confirming agreements of promises is to send a memorandum of thanks in which the director details steps already taken to implement the agreements reached. As, for instance, "Thank you very much for. . . . The staff were delighted when I told them. We have notified all the subclub chairpersons and they are. . . ." The more persons that have been involved and the further implementation has proceeded, the less the chance that the director's version will ever be questioned.

As a program director, however, never try to take advantage of the suggested method dishonestly or incorrectly. To attempt to push the management representative one step beyond what was in fact agreed upon may be disastrous.

Memoranda of clarification and confirmation are important tools of communication at all levels. They are the best way to ascertain that there really was a meeting of the minds. They also guard against the frailty of forgetfulness. They guard against future misunderstandings and conflicts. If the matter discussed is of significant importance, the director should initiate such memoranda after meetings with subordinates also.

Minutes of meetings serve the same purpose. For that reason minutes should always be kept of all staff meetings and all committee meetings. Not seeing to it that such minutes are kept is a common failing of administrators. Even at the moment of occurrence, communication can never be perfect. With the lapse of time, it is even more imperfect.

The art and skill of effective communication is of the highest importance to recreation program directors. Yet on at least an equal plane of importance is the art and skill of relating to others, which is the essence of personnel administration. In this area the program director continues the role of leadership and the role of chief communicator, but adds the role of administrator.

Personnel Administration

Personnel administration is one of the most complex and demanding tasks of administration (which is why there are whole books and courses about it). Just as there are no two persons exactly alike, there is no person who is always the same. Dealing with personalities is like trying to hold quicksilver which divides and combines and rolls around with no forewarning. Although the fact may not always be recognized, the program director must extend personnel administration beyond the paid staff to all the volunteers, which does not make the task appreciably easier.

The Value of Volunteers

The wise director, regardless of the size of the paid staff, never attempts to run a recreation program without a sizable number of volunteers. It is the volunteers who identify the program as an employee program. The volunteers serve as the cement which holds the organization together and makes it go. The volunteers communicate the desires and attitudes of the membership at large. Each involves others. The larger the number of volunteers, the greater the overall involvement and commitment of all members. Only the volunteers make it possible to achieve the ideal of a program "of, by, and for" the members.

Nevertheless, volunteers are not altogether assets. They, or some of them at least some of the time, appear on the other side of the ledger as liabilities. Volunteers require more direction and development than the paid staff. For many, it may be their first experience in a leadership role. Some of them demand more of the director's time and attention than they contribute to the welfare of the program.

Assuredly there will be times when the director will be able to look back and know that it would have been easier to have done the job personally than to have had it done by volunteers. Whether a given volunteer is an asset or a liability depends somewhat upon the character and the ability of the volunteer, but probably depends even more upon the quality and amount of training, supervision, and guidance provided by the director.

The Selection of Volunteers

Volunteers range from the members of the governing board (unless it is a "company" program and the board consists of Management representatives) through the executive officers to the chairperson of the smallest subclub. They are selected in different ways from one association to another and even within the same association. In few instances, however, are they volunteers in the literal sense, that is, persons who, figuratively speaking, took two steps forward and said, "I will." Too often persons volunteering in that sense are burning with enthusiasm one week but burned out the next week, unable or unwilling to carry on with the responsibilities so lightly accepted. Consequently most volunteers are either selected (hand-picked) or elected (group-picked).

If volunteers are selected, the director should have an important voice in the decision, but not be alone. Either the volunteers should be recommended by the director, subject to the approval of the governing board, or the recommendations of the association president should be approved by the director before any announcement is made. The former method is usually superior because over the course of time the director is best able to identify those who are most able, dependable, and valuable. Yet the director should be careful that there is a good mix of the experienced and the untried. Only so can the function of the development of leadership be fulfilled.

If, on the other hand, volunteers are determined through election, the program director should avoid the temptation to exert even the slightest suggestion of influence on the process. In that case, the intervention of the director is both unwise and unnecessary. The democratic process results in some poor choices, but so does the hand-picked process, no matter how able and careful is the director. In the long run the democratic process achieves surprisingly good results. Most important is what happens after the selection or election as the case may be.

Committees

Of all volunteers, committees are those most often misused and abused. Someone once defined a committee as "the unwilling, appointed by the unable, to do the unnecessary." That such a "definition" should have been concocted and that it strikes one as humorous is a sad commentary upon the ability of many an administrator.

The first rule should be that a committee is never appointed unless there is something worthwhile for that committee to do. Too often, administrators, having encountered some opposition to a new policy or decision, arrange for a committee for the sole purpose of endorsing the decision or policy desired. Or, an administrator appoints the "right" persons who are then led to recommend the policy the administrator secretly desires but has felt it unwise to announce directly. The use of "rubber-stamp" committees will invariably be discovered in time, to the detriment of the program director and the program. Those members so used become resentful, and the director loses credibility with all members who become aware of the practice.

Rule two is that each committee should be presented with a written charge. That charge defines the task which the committee is asked to accomplish, specifies the parameters within which it is to work, clarifies the authority of the committee, and establishes the time frame within which it is asked to work. Oral or vague instructions almost guarantee unsatisfactory results or a lot of wheel-spinning while the committee discovers what should have been specified in the beginning. Committees are often a special challenge to program directors. They call forth one of the elements most necessary for a good administrator: patience.

Rule three is that adequate time must be allowed for committee deliberations. Committees seldom unerringly and quickly find the pathway to the objective. It takes time for committee organization, for the identification of those members who will be most productive, for the assignment of responsibilities, for the assimilation of ideas, for the exploration of several pathways, and for the often lengthy discussions which lead to a final concensus. During that time, it is legitimate and sensible for the program director to make subtle suggestions which the committee later may, or may not, adopt as its own.

Rule four is that the results of committee deliberations are to be taken seriously. In no case should a committee report be quietly slid under the table in the hopes that it will be forgotten. If the report is a recommendation, that recommendation should almost always be followed. If it is not, the committee is entitled to a full explanation. If the committee was empowered to make a decision, the decision should be implemented. Otherwise capable administrators have hastened the end of their careers by a cavalier attitude toward their own committees. Volunteers, whether in committees or singly, deserve to be taken seriously and treated with respect.

Other Volunteers

Success in working with volunteers depends largely upon four elements: adequate written guidelines, sufficient introductory training, continuing supervision, and reasonable reward—usually recognition.

All volunteers, from highest to lowest, need to know what is expected of them. For each there needs to be a job description. If such a description is not provided in the association bylaws, it needs to be included in a policies and procedures manual—described later. Volunteers need to know what are their responsibilities and what is the extent of their authority. Each needs also to know the accepted association policies and procedures.

There also need to be special officers' clinics or workshops. These are to enlarge upon the policies manual, to answer unanticipated questions, to clarify misconceptions, and to make the newcomers feel comfortable in their roles. Such sessions also place the newcomers in touch with each other and establish a more familiar and comfortable relationship with the director and staff. Yet written guidelines and orientation sessions are only the beginning.

The volunteer worker needs and deserves continuing supervision. Good supervision is a positive quality. It is never designed to catch people in errors or failures; its aim is to facilitate success. It builds upon strengths and improves upon weaknesses. It provides knowledgeable, tactful and well-meaning, but honest feedback. It is a ready source of advice and discreet counseling. Effective orientation, a well-written service manual, and effective supervision should lead to success; and success should meet with recognition.

Public recognition for volunteers should not be perfunctory but earned, deserved, and then be given graciously and honestly. The kind and amount of recognition should be in keeping with the importance of the accomplishment. On some occasions the contribution will be such that the director should arrange for a high ranking company or public official to make a public award.

Beyond Recognition

A natural capacity, developed and tested in the employee activity program, should be brought to the attention of the company management. The program director who makes the qualities of such a member known to management serves two important purposes. First, since leadership is a valuable commodity, the deserving member may be promoted within the company. Second, the company will discover that the employee association is a free and reliable training ground for company leaders. Once having made that discovery, the company will avail itself of that asset with greater frequency, to the mutual benefit of the company, the individuals so discovered, and the employee association.

Sometimes the recognition of leadership qualities of employee association members will occur fortuitously as the different levels of workers associate in recreation activities. It will happen much more often if it is a deliberate objective of an able program director. The director should be on the lookout for qualities such as personal integrity, the capacity for handling details, the ability to inspire followers, the willingness to accept responsibility, and a flare for originality and initiative.

On each occasion when the director is made aware of such qualities of volunteers in specific situations, a record should be made. Recommendations which mention specific examples carry much more weight. Negative observations, incidentally, are both unnecessary and unwise to put into writing. The director will be reminded of weaknesses by omissions and will write recommendations accordingly. Experienced administrators and personnel officers judge recommendations both by their inclusions and omissions. The program director who makes only positive and completely honest comments is fair to all involved. The program director has a professional responsibility to seek the personal development and the ethical promotion of all fellow workers, whether volunteers or members of the full-time staff. To the latter we now give attention.

The Full-time Staff

In contrast to the volunteers, the full-time staff is relatively much more stable. There are no annual, large-scale turnovers, and additions are usually one at a time. The full-time staff composition in large recreation associations is extremely varied. It may include professional assistants, semiprofessional technicians, interns, clerical personnel, and maintenance personnel. These are very diverse in their interests and in their needs.

However large or small the full-time staff may be, there is one overriding consideration: every single member needs to be made to feel a member of the team. This probably is true of all organizations; it is particularly relevant for employee activity associations. The success of the organization rests upon the extent of the esprit de corps of employees and members alike. The implications for personnel management are many.

Every employee needs to be familiar with the purposes and objectives of the entire organization. Every employee needs to be committed to those objectives and understand how the personal area of responsibility contributes to those objectives. More than that, each employee needs to have some idea of the responsibilities of every other personnel position and how each position contributes to the general goal.

The extent to which there is a genuine personnel team depends largely upon the leadership. The director sets the example of treating each employee and each position with respect. The director talks *with* and not *down* to even the lowest ranking employee. The director's attitudes and actions set the tone for all other employees.

There should be some meetings of the entire staff. At least some times all the members and their spouses should meet socially and informally. Whenever there is a meeting of the director's staff and the consideration of a special operation, as for instance the camping facility, all those concerned with that operation should be invited to attend. Those nonprofessional persons often are possessed of insights that might otherwise be overlooked, and they often have valuable suggestions to offer. Beyond that those persons are much more likely to fully support any decision made in respect to the operation in which they are engaged.

Circumstances do alter cases, especially as to emphasis, but it is surprising how many of the practices rated as important for the training and development of the professional staff apply also to all other employees. The director should arrange for well-planned, in-service training sessions and seasonal "refresher" orientations. These should be led by someone who has special qualifications in the area(s) emphasized. That person may come from the business community, near-by university, or be a member of the regular staff.

As feasible, there should be rotation of areas of activity and responsibility.

The director should take care to delegate responsibility and authority. Most persons grow as they need to.

The director should teach by example. When feasible, for instance the director should take someone of the staff along on a mission to negotiate a new buying contract or to discuss a proposal with Management.

Finally, the program director should consider it a professional responsibility to work individually with each association employee toward growth and development. Each individual has different areas of strength upon which to build and other areas which need strengthening. Sometimes the avenue may be personal counseling; sometimes it may be provision for different experiences; sometimes, suggestions of books or articles for reading; and sometimes, formal courses of study. Always the approach is nonthreatening, positive, and ego-building.

A Program Example

Combining the theoretical and the practical, the following excerpt is one part of the carefully developed leadership development program of the Kodak Park Activity Association in Rochester, N.Y. The K.P.A.A. full-time staff consists of the executive director, eight staff personnel, eight bowling lane assistants, and an office force of six. Its staff development policy is quoted verbatim.

K.P.A.A. Professional Staff Development[1]

Almost invariably new professionals must be trained; experience for this position must be developed on the job. The executive secretary assumes responsibility for the training of new personnel with assistance provided by the more experienced staff personnel. Training is provided in the following ways:

1. Learning on the job—new personnel are given assignments of lesser difficulty to begin with. The initiative to develop and advance to more difficult tasks varies with the individual and is a determining factor as to when more difficult responsibilities are assigned.
2. Rotation of assignments—once assignments of lesser difficulty are mastered, new personnel are put in the situation of a competitive and challenging rotation of assignments with other staff people. Assignments in the various clubs and activities are rotated every two or three years. This procedure prevents stagnation and encourages growth. Over a period of time each activity or assignment has been handled by many different staff people with different capabilities. . . . Staff people are thus challenged by change and competition.
3. Special assignments take on many phases:
   A. Special courses—many courses are available through the company and offer the individual the opportunity to better understand and develop techniques that can enhance job performance.

    B. Leadership assignments—assignments involving the coordination of other staff members or involving the department give the staff person an opportunity to develop.

    C. Special reports—offer the staff person an opportunity to deepen the understanding of the program in general or some detailed phase of the program.

    D. Serving as delegate to a local or a national meeting or convention offers a broadening experience with the person learning more about the scope of the profession.

    E. Serving on staff committees which investigate or report on some phase of the program or which solve a problem.

    F. Serving as the executive secretary in his absence.

    G. Serving as a staff representative to a Board of Governors' committee gives the staff person experience in the educating and guidance function.

4. Staff meetings—staff meetings provide a medium to give direction, pass on information, identify and solve problems, discuss situations, brainstorm, and decide in a group situation. They can increase morale and kindle interest. They provide valuable experience to staff people for working and leading in the group situation.

5. Personal assistance in individual development—this involves the person-to-person consultations with the executive secretary. This approach is perhaps the most effective in terms of developing potential. This method implies to the staff person a sense of worth and personal interest on the part of the supervisor. This method also allows the supervisor to advise, counsel, and guide professionals on a one-to-one basis.

Organizing for Efficiency

Most of this chapter has been devoted to the special kinds of responsibilities required of the leadership role of the program director. We turn briefly now to four specific administrative tasks which, taken together, contribute significantly to the efficient operation of an employee activity association. These responsibilities are: policies and procedures manuals, budgets and fiscal controls, annual reports, and evaluations.

Policies and Procedures Manuals

Policies and procedures manuals are the single most effective means of achieving efficiency in employee activity associations, characterized as they are by periodic changes of persons in leadership roles. Certainly there should be well-planned orientation meetings and officers' clinics, but these must be supplemented by manuals. We cannot expect persons to remember the details presented orally. Even those leaders of experience need manuals to refresh their memories about procedures which may seldom have been used. Manuals help to eliminate confusion, mistakes, and omissions.

Policies and procedures manuals are so important that we are willing to state categorically that no large recreation association can operate smoothly and efficiently without a manual. Even the smaller organizations definitely should have a manual. Just possessing something called a manual is not enough. The manual needs to be sufficiently inclusive to answer the vast majority, perhaps 90%, of the questions that arise from day to day in respect to policies and procedures. The original manual prepared for an organization may not be anywhere near that inclusive, but the manuals should be constantly improved in use.

Regardless of how careful the preparation, every new policies and procedures manual will contain some errors. It will almost certainly contain some passages which are confusing. Actually, it should be a matter of congratulations if there are not some internal contradictions. In addition, there are inevitably some changes which make one or more sections of the manual obsolete before a year has passed. These changes may stem from a revision of the bylaws, decisions made by the board of governors, changed Company policies, or even changes outside the Association and the Company.

In order to facilitate the correction of errors, the addition of unplanned omissions, and the inevitable changes, policies and procedures manuals should be loose-leaf. Thus changes can be made as needed by the substitution of a page or two. Otherwise it is impractical to make changes more often than once a year, and even annual reprints of a reasonably inclusive manual are quite expensive. Another reasonable approach is to print the manual in small separate sections according to the areas included. All that is important, however, is that there be a manual, constantly improved and periodically updated.

Manuals are particularly valuable in respect to fiscal controls. It seems that a great many persons placed in a position of responsibility in an area which has a budget and funds do not anticipate any controls. They appear to feel that they should be able to order what ever is useful and make charges or write checks as long as the money lasts. One of the most important inclusions in manuals, then, is a section which is quite specific about expenditures: under what circumstances purchases can be made, whose authority is required, what are the necessary steps in requisitioning and making payments, and what records must be kept. The manual should include specific instructions in these matters and sample copies of the forms required along with illustrations of how they are to be completed. Nothing must be left to the imagination.

Among other important inclusions are job descriptions for all leaders, full-time and volunteer. These should be sufficiently detailed that the area of operation, the responsibilities, and the authorizations of each position are clear to the holder of each position and to everyone else. Confusion, conflicts, and disputes are thereby minimized; and responsibilities are more likely to be met.

As those who have prepared an original manual of policies and procedures know, manuals require a large expenditure of time and painstaking effort. The job of keeping them updated is tedious and unending. They are the special responsibility of the program director who may delegate some of the work but whose personal involvement is essential. A good manual is one of the distinct marks of a professionally directed program.

The need for thorough policies and procedures manuals is well-recognized by the professional industrial recreation administrators, and several excellent examples exist. We are especially impressed by the "O-I Building, OnIzed Club Guide Book" (Owens-Illinois, Toledo, Ohio) originally developed by Mel Byers and meticulously revised and updated by Martha Byers, both CIRA. The 1973 edition, of which we have a copy, runs to 97 valuable pages and has an additional and thorough four-page index.

The section respecting the disbursement of funds includes blank forms and sample executions for the transfer of funds, deposit of funds, check authorization, club requisitions, records of payments of dues, and membership data records.

It includes the names of the current board of governors and of the officers of the special interest clubs as well as the names of all the current committees. Copies are placed in the hands of all officers and other copies are kept available for the reference of all members. It is attractively but inexpensively bound, loose leaf, and updated regularly. Much of its contents can be inferred from its Table of Contents.

TABLE OF CONTENTS

Budgeting

Budgeting is another administrative area where the responsibility falls upon the program director. Budgets force organizations to look ahead, establish priorities, and allocate scarce resources. So-called annual budgets usually include two different kinds of budgets and two different time periods. The operating budget will match income with the needs of the association for the coming fiscal year. The capital expenditures budget makes provision for the costlier acquisitions over a longer period, perhaps five years. The capital expenditures budget is necessary because, if the association plans expensive construction or renovation in future years, it usually is essential to plan to put aside some of the funds from each of the intervening annual budgets.

There are many avenues to budget-making. If the employee activity program is a "company" program, the company will usually inform the program director how much money the company will supply for the coming year. All that is left is for the program director and/or the governing board to allocate the money between the various activities. In some instances the company, perhaps through a company-dominated governing board, also determines the allocations though suggestions, or requests may be asked from the different clubs and activities.

At the other extreme is the employee association which raises a great deal of its own funds. In this case the budget requires a careful estimate of the funds expected to be available as well as a plan for their expenditure. Although the final decisions will probably rest with the board of directors, the board will almost certainly expect the program director to do all the work. It will look to the director for information regarding the historical record of allocations as well as for new developments which may have a bearing upon next year's income and expenses. Usually after the director has made an estimate, which sometimes is more of a "guestimate," the board will select the income figure to be used for the budget, but, more often than not, it will be the amount recommended by the director.

Thereafter the board, perhaps through the director, will request the various departments and clubs to submit their budget requests for the coming year. Another fairly common approach is for the board to set aside a certain amount for new programs and activities plus a sum for contingencies, then prorate the balance among the various funded divisions on a historical basis. That is, the amount proposed for each division will be related to the amount allocated and spent by that division during the current year.

Another and less common approach is what is called "zero-based" budgeting. In that approach, which is more difficult and complex, each budgeted division is supposed to state its objectives for the coming year. The division then is asked to present a rationale for each item or class of items needed to meet its stated objectives. This does force each division to focus upon objectives. In theory at least, this method rewards the dynamic and active budget divisions and penalizes older but declining divisions which otherwise might continue for a long time to coast upon past performances and larger memberships.

Zero-based budgeting has one serious drawback. It gives a tremendous advantage to the divisions whose leaders are the most imaginative and aggressive. Those leaders tend to ask for extremely large funds and imaginatively come up with persuasive rationales. What happens too often is that the governing board is mislead. Upon finding that the amounts requested and "justified" exceed the expected income by a substantial amount, say 25%, the board may solve the problem by deciding that each division shall have its budget reduced by 25%. The divisions which requested the imaginative and aggressive budget allocations, will still be in excellent shape after the 25% reduction. But those divisions which submitted honest and realistic budgets will be in real trouble.

If zero-based budgeting is employed, it is imperative that someone having a really good view of the entire operation is able to spot the exaggerated budget requests so that reductions may be in line with the reasonable needs of each division. The only person able to do that will be the program director, who is then very much on the spot. This is another situation when careful homework is important. The director then is in a position to furnish all the relevant information to the board of governors. The director should also be prepared, if asked, to make a recommendation; but the decision should be made by the board.

Actually, no matter what the budget process, the role of the director should be to gather and present information and, if desired, make recommendations. The role of the board is to make decisions. People are never more "touchy" and emotional than when money is concerned, and the board of directors as a collective body will invoke much less resentment and ill will than will any single individual.

The preparation of budgets entails a lot of work for the program director under any circumstances, but the task will be a great deal less onerous if careful records have been kept, the same kind of records which will be useful for the annual report.

The Annual Report

The organizational annual report is the inescapable responsibility of the program director. The director may delegate responsibility for parts of the report but is personally responsible that the report is prepared and is personally responsible for its contents.

Almost any factual annual report is better than none. Some are not more than a statement of financial affairs: expenditures, income, current financial position, and financial planning for the coming year. Both the interest and the value of the report will be improved by the addition of participation records. How many attended each activity? How many availed themselves of each service? A simple method of further enhancing interest and acceptability is the addition of a few well-chosen pictures. Very effective are activity group pictures showing individuals. Good annual reports will also include a well-written narrative, a running commentary upon the data and all other aspects of the report. The commentary will aid in the correct interpretation of the data and point out the essential importance of data that might otherwise be overlooked. The narrative may carry the document beyond a mere quantitative report to an evaluative study. It is well worth the extra effort.

Values and Quality

There is a terrific difference between the value of a so-so annual report and a very good one. We believe that the quality of annual reports will be better if the benefits of a good annual report are kept in mind. Accordingly we list a few.

1. It will be of much assistance in selling Management on the value of the program. Here are presented in one place, accomplishments over an entire year. They are the more impressive when listed together.
2. It will generate greater enthusiasm. Those who have participated extensively are reminded of the good things they have shared. Those who have not, may realize what they missed.
3. It will serve as a valuable educational device for new leaders. It is of especial value for new members of the board of governors.
4. It is of much assistance in future planning. It helps in the avoidance of past errors and serves as a reminder of successful practices. The report, sort of a bird's-eye view of the year, gives a clearer perception both of direction and progress.
5. It helps everyone, Management and association members, to appreciate the contributions of the program director. Some of those persons never realize and others tend to forget the accomplishments of the program director. The annual report is therefore both a justification for the position of program director and for the salary of the incumbent.

Yet in spite of all the values of a good annual report, it should be obvious that the annual report has no miraculous qualities. It cannot make a poor program into a good one. It cannot transform a year of little achievement into one of glowing accomplishments. It will certainly be a matter of great embarrassment if, when the time comes to prepare the annual report, the director realizes that in truth little has been accomplished. One means of avoiding that untimely discovery is to arrange for ongoing program evaluation. Such evaluation also has other important benefits and deserves attention all aside from the annual report.

Program Evaluation

In general, there are two basic kinds of evaluation. One is ongoing, informal, and internal. The other is periodic, formal, and external. Let us begin with the internal process.

Informal evaluation goes on all the time as the program director and others form impressions about program activities and services. If, however, those impressions are aimless, purely subjective, and unrecorded, they will be of little material benefit, and they probably soon will be forgotten.

Deliberate informal evaluation is undertaken for the purposes of planning and decision making. The program director needs to arrange that there is timely evaluation of every activity, service, and project. The director can make some evaluation personally upon the basis of observation and participation and cost figures. All these should be made a matter of written record.

Also, at the conclusion of each major event or project, the director should sit down with the staff and/or leaders involved and discuss the major achievements as well as the kinds of problems encountered. These should be recorded in order to build upon strengths and avoid the repetition of mistakes.

It is also very valuable to obtain the reactions of all the participants directly involved. At the end of a season or major event, the director or staff member might discuss the outcomes with the subclub membership or other participants.

Questions are: What things went especially well this time? and What should we do differently next time around? Though there may be differences of opinions, out of a group discussion some concensus will probably emerge. Let it be recorded.

Another useful method is a questionnaire, completed, desirably, on the spot. Such questionnaires need to be brief and reasonably specific. They should be carefully prepared because if vague or overlong they will be sloppily completed if at all. The more specific the questionnaire, the easier its completion will be. Desirably participants should be able to indicate their responses with a simple checkmark. This also permits tabulation and analysis. At the end of each questionnaire, there should be a space left for comments. Some of the comments may be enlightening and reveal aspects which the sponsors might otherwise not consider. Let us repeat, the results should be made a matter of record.

Formal Program Evaluations

This second type of evaluation is rather rare in industrial recreation at this time. The periodic, formal, and external evaluations should be encouraged, for they are more objective and more revealing. They are more objective because conducted by outsiders who have no personal axe to grind. Also they should be conducted by persons who have special training, experience, and expertise in such procedures. In the formal evaluation, the participation of the program director should be strictly limited. The professional and able program director will initiate the formal evaluation, make the preliminary arrangements, and facilitate and support the efforts of the evaluators. Thus the director will create a favorable climate for the evaluation in order that the staff and all members will be receptive and cooperative. Beyond that, the director will see to it that helpful material is collected in advance of the evaluation visitation and will provide answers for unanticipated questions asked by the evaluator(s). Only thus can a reasonable degree of objectivity be achieved.

Formal program evaluations need not be frequent. One in an eight-year period is probably adequate. Formal evaluations require a great deal of preparatory time and effort on the part of the program director and staff as well as on the part of volunteer leaders. They are also expensive usually. The amount of the expense will depend upon such factors as the extent of the evaluation and the size and professional standing of the evaluative team.

The best and most helpful evaluations will be conducted by a team which will probably include one or more university professors teaching in the field and who have special training in evaluation techniques such as surveying, sampling, interviewing, the preparation of questionnaires, and the analysis and interpretation of findings. Other members of the team should include esteemed and experienced industrial recreation program directors from outside the immediate geographical area. The participation of some member of the national NIRA staff would also be an asset. Depending upon the complexity and emphases of the program to be evaluated, other additions to the evaluative team may include experts in such special areas as, for instance, physical fitness. Obviously the services of such a team, even for three or four days, will be expensive.

The best and most useful evaluations will be concerned with the total program: program activities and scheduling, services, facilities and equipment, and the administration, *including the program director*. Unfortunately, some persons consider such a complete evaluation threatening, and that may be part of the reason that so few such evaluations occur. In fact, professionally led evaluation teams come with only one purpose in mind, and that is to be helpful. They most certainly do not come to be destructive nor to hunt scalps. The reports, oral and written, will emphasize strengths; and whatever criticism is included, will be *constructive* criticism.

Less valuable, yet still very worthwhile, are formal evaluations that are less extensive and require fewer evaluators, perhaps only one or two. Even an evaluation limited to the adequacy of programs and services is extremely helpful and should be undertaken if more extensive evaluation seems inexpedient. A program director who desires such a limited but professional evaluation would do well to get in touch with the nearest university which offers related training. Some professor will almost certainly be willing to conduct the evaluation. Also, much of the time, the professor will be able to secure assistance from senior program majors or graduate students who will be happy to gain the experience and added perspective.

Also through the medium of a nearby university, free evaluations may be obtained. These may be provided by graduate students who desire experience and may wish to use the acquired information (anonymously if the program perfers not to be identified) as the basis of a masters thesis. The benefits from such student evaluations are quite uneven, depending upon the seriousness and ability of the students and the quality and extent of supervision by the major professor. The results are sometimes quite perceptive and most helpful, as the literature in the field demonstrates. On the other hand, if the evaluation is not well done, that fact is usually readily apparent in the report, which can be treated accordingly. The strong chances are that such evaluations will be useful, both to the program evaluated and to the students concerned; and such liaison projects with universities are to be encouraged. Through whatever media, it is almost always helpful for us all to see ourselves as we may be seen by trained and objective observers.

Conclusion

With program evaluation, we conclude this chapter concerning the responsibilities of the program director, directly or through the staff. The treatment was not intended to be all-inclusive nor definitive. In every area mentioned, the industrial recreation professional will need more extensive and intensive preparation. Professionalism in industrial recreation—the meaning of and the avenues to—are the concerns of the next chapter.

Notes

1. Courtesy of Kirt Compton, CIRA, Executive Secretary of the Kodak Park Activities Association. He also sent other materials for this book.

PART III

9
Professionalization

OP MANAGEMENT SPEAKS

The industrial recreation program at Danly has survived and thrived since the earliest days of the company. The word "survived" is used here because we, as a company, have no recreational area, no recreation facility, not even an assembly room. Recreation has thrived, despite these drawbacks, because we have an employee population which is interested in group recreation, a Recreation Director to assist the various employee committees in their planning, and a company philosophy of encouragement and cooperation.

There is outside competition in baseball, basketball, golf, and bowling. But the truly rewarding programs, the real morale supporters, are intraplant. The strength of the desire and the need for industrial recreation is best illustrated by the fact that all the groups except our duplicate bridge group which meets in the company cafeteria, must find outside facilities for their activities.

The programs are diverse and include the Sportmen's Club with hunting on its own preserve, fishing contests, leagues in the popular sports, and planned tours to foreign lands. Our golf committee has developed a competitive schedule very like that of a private country club with many interesting weekend events in addition to the after work league play.

James C. Danly

Also Pg. 177

We stand ready to support any worthwhile activity. The size of the interested group is not important. If the idea is a good one participation will inevitably increase.

James C. Danly
President
Danly Machine Corporation

The professionalization of Industrial Recreation is of great importance to the future of that field and to all those affected by it. That includes students who see a future in it for themselves, present practitioners, and industrial leaders who recognize its potential. The subject is difficult because the word professional has a variety of meanings in use. Sometimes professional merely means paid as opposed to volunteer or amateur. While the word professional in that sense is too limited to be satisfactory for our purposes, it does suggest two ideas that we would apply to professionalism in the larger sense. These are: (1) the professional has special expertise (Why else would they be paid?) and (2) the professional considers that activity important enough to deserve full-time attention. Thus we eliminate volunteers and hobbyists from our concern and concentrate on those who earn their living by the professional activity.

The term professional becomes more meaningful and useful if we limit it to those persons who conform to the standards of a profession. This does not completely eliminate confusion because the professions are not sharply distinguished from other occupations. Further clarification emerges if we limit professions to those occupations which provide highly specialized, intellectual services. Yet that does not resolve all the difficulty.

The number of highly specialized, intellectual occupations is increasing as new fields of intellectual endeavor develop in response to the increasingly scientific and sophisticated technology of the world in which we live. Such fields may most often be identified by the extensive courses of training which are prerequisite for employment in those fields. Yet educational requirements do not of themselves define a professional occupation.

At one time there were only the three learned professions: theology, law, and medicine. Today there are in the neighborhood of ten times that number of fields of specialized and extensive learning, and it is likely that no two persons would be in complete agreement as to which of them constitute a profession. In terms of public recognition as a profession, perhaps a dozen and a half new occupations have clearly arrived; but there are at least that many more in the process of development and striving for public acceptance and recognition. Among these latter is Industrial Recreation Administration.

How far industrial recreation has come along the path of public recognition and acceptance as a profession may best be judged against the background of the highest attributes of professional occupations. There are six generally accepted characteristics of professions.

1. There is a standard of success related more to the service of society than to personal gain. This does not suggest that the motive of personal gain is absent. It means that professionals do not consider their careers successful unless they have made a significant contribution to the general social welfare. One of the clearest indicators of professionalism is the willingness, even the compulsion to forego personal ease, personal pleasure, and personal welfare to the needs of others in the practice of the profession.

2. There is a body of erudite, specialized knowledge which is essential to the occupation. Thus formal study at an advanced level is essential to a profession. In the beginning there had to be a small number of gifted leaders dedicated to inquiry, learning, and progress based upon experience and experimentation. Once fully established, the only access to a profession is formal study. It is true that, here and there, a few gifted individuals might be self-taught and possess the necessary body of specialized knowledge; but the problem is that there is no way of identifying those individuals or of evaluating the depth of their specialized knowledge. As a practical matter, then the only avenue to a profession must be formal study.

3. There is a recognized set of attitudes and ethical standards for the application of the specialized knowledge. These attitudes and ethical standards are, in a sense, a professional conscience—personal/professional values and commitments akin to the Hippocratic Oath in the profession of Medicine.
4. There is an emphasis upon lifelong growth in knowledge, skills, and in the ability to be of service. The professional is always striving for improvement, alert to change, and receptive to new professional discoveries.
5. There is an association of practitioners. That association is dedicated to the advance of the profession, the welfare of the public as it is affected by the profession, and to the maintenance of professional standards.
6. There are standards that govern admission to the practice of the profession. These constitute a means of identifying those who are qualified in terms of the specialized knowledge, the essential skills, and all the other criteria of professionalism. Such standards have to be developed by the profession itself, though, through professional practices acts, the State may regulate the right to practice. Government licensure of practitioners is not, however, an indication of professionalism of itself since many state and local governments regulate and issue licenses of many trades not even remotely professional.

Industrial Recreation as a Profession

Upon the background of the recognized attributed of the long-established and recognized professions, it is obvious that Industrial Recreation Administration has come a long way, but has some distance to go before it achieves public recognition as a profession. We are not speaking of individuals; for, among the current industrial recreation administrators, there are many who meet all the criteria of professionialism. Yet some do not, and that is a problem. Another serious problem is that the public in general does not recognize the professionialism of those individuals who are true professionals. The ultimate goal which must be achieved—and will be—is widespread public recognition of Industrial Recreation Administration as a profession. Before pursuing that topic further, it may be worthwhile to consider the situation in respect to recreation as a profession.

Recreation as a Profession

Though there have been gains in recent years, the field of recreation as a genre has not yet achieved widespread recognition as a serious, respectable, and professional field of employment. In recent years a great many colleges and universities have added departments of recreation or have combined recreation with existing departments. In the academic world, recreation is coming to be recognized as a special professional field. Yet most of the public at large still conceives recreation as only fun and games or athletics. Probably the majority of laymen still suppose that anyone engaged in recreation as an occupation is an athletic coach, as, indeed, most of them were only a decade or so ago. The field of recreation does include coaches, of course, but it also includes a great many others who are neither qualified for nor interested in coaching as a career.

Apparently we have reached the stage where at least the better educated and informed public understands that recreation also includes aesthetic, cultural, and educational experiences and growth. Some know that recreation today is devoted to the development of the whole person and the enrichment of individual and community life. To that extent we may say that there is limited public recognition of recreation as a profession, though even that is not of much help to the specialized field of industrial recreation.

The Uniqueness of Industrial Recreation

The field of industrial recreation is almost unrecognized even in academia. College professors in general have almost no idea of what it is. Even among those professors teaching the field of recreation, relatively few know what industrial recreation is (Most of them suppose that it is recreation in an industrial setting). What is worse, some of those presently earning their living in the field of industrial recreation are not very clear as to what is so unique about it.

The Enlightened

And who are the enlightened ones? In addition to you, who have read Chapter 3, they include a great and growing number of dedicated and progressive industrial recreation administrators and a significant and growing number of business and industrial leaders who recognize the potential of industrial recreation.

The enlightened ones see the field of industrial recreation as recreation in its broadest sense plus a great number of other activities and services provided for the employees of participating businesses and industries. The leaders in the field of industrial recreation sense the partnership of industrial recreation and the sponsoring company. They see it contributing to the goals and objectives of the company in the interest of society. They and the industrial leaders of the kind represented in *Top Management Speaks* see industrial recreation as going hand-in-hand with the growing acceptance of the social responsibility of business. They see it as a unifying force between business, labor, and the social community. They see it as a program which emphasizes a nonadversarial relationship between employee and employer on the one hand and which contributes to the welfare of both and to the community at large. No one expects industrial recreation to bring about millenium, but the enlightened believers see that industrial recreation is a program of social significance.

Milestones Toward Professionalism

The professionalization of any occupational field takes years and requires the unremitting effort of those leaders with vision who are true professionals before their time. Industrial recreation has its roots in the nineteenth century, but it did not acquire anything like its present form and substance until far into this century. For a long time, no one appears to have thought of it as a full-time occupational field, much less consider it as a profession. We would date the visible beginnings of the desire and efforts to professionalize Industrial Recreation Administration at 1938, four decades ago.

We cannot know when the concept of professionialization came into the mind of one man and was discussed with others who became caught up in the enthusiasm and became part of the effort. We cannot know of the countless discussions of small groups of leaders nor of all the developments which contributed toward the professionalization of Industrial Recreation Administration. We can identify bench marks when the ongoing efforts became visible through significant developments. These milestones we do note, though it must be remembered that during all the intervening years the field of industrial recreation was gradually changing and the slow progress toward professionalization was taking place. Let us look at the visible milestones.

1938

Purdue University, under the leadership and direction of Professor Floyd R. Eastwood, established a special program for the education of industrial recreation administrators. Thus, at least Eastwood and some others, recognized that the administration of industrial recreation encompasses a body of specialized and erudite knowledge, one of the important criteria of professionalism. Like many another visionary, however, Eastwood was too far out in front of his contemporaries. At that time, not enough industrial leaders or employees interested in industrial recreation understood as did Eastwood that there was the specialized body of knowledge for Industrial Recreation Administration. Hence there were not enough jobs available for the students who did come into the Purdue program. As a consequence, the number of students in the program dwindled until it was no longer feasible for the program to be offered. The first effort toward professionalism in the field was aborted, but not before much preliminary research had been done by Eastwood's students. Undoubtedly the original Purdue program measurably advanced the professionalism of the field.

1941

A few farsighted leaders, notably influenced by Floyd Eastwood and financially supported by some half-dozen industries, met and formed the Recreation Association for American Industry with Floyd Eastwood as its first president. With only a change of name—to the National Industrial Recreation Association—that association proved to be permanent, thus fulfilling another of the six criteria for professionalism.

1955

The first textbook for Industrial Recreation was published by the McGraw Hill Book Company, Inc. Written by Dr. Jackson M. Anderson, formerly a professor at Purdue University and then Director of Research for NIRA, it provided further evidence of the existence of a field of specialized knowledge. That book, which will always be a landmark: *Industrial Recreation: A Guide to Its Organization and Administration,* was excellently written. Incorporating much of the research done by graduate students in the Industrial Recreation program at Purdue, Anderson's book was *the* authority in industrial recreation until time gradually eroded its relevance. That that excellent book was never revised is probably explained by the fact that, like the Purdue University Industrial Recreation Program, it was ahead of its time. The professionalization of Industrial Recreation Administration—to which it made a significant contribution—still had not made sufficient progress. There were not then sufficient students of industrial recreation to make a revision and a new publication economically feasible.

The chief difference between this book and Professor Anderson's excellent first text is not a matter of timeliness. Dr. Anderson's book intended to, and to a remarkable extent did, summarize vitually all that was known about the field of industrial recreation. In the intervening quarter of a century, the field of industrial recreation and the specialized knowledge about it has grown so voluminously that it is no longer feasible to consider summarizing all about it in one volume. Rather than including the field in its entirety, as Professor Anderson ably did, this book is intended to serve only as an introduction to the field. Other textbooks will be written for special courses in Industrial Recreation Administration, Industrial Recreation Supervision, et cetera.

1958

NIRA began publication of *Recreation Management*. This publication represents a tremendous acceleration of the growth of professionalism. From its very beginning, it has served as the chief instrument for the professional growth and continuing education of industrial recreation administrators. *Recreation Management* today, however, is a far cry from what it was in 1958. Anyone interested in the evidence of the gradual evolution of industrial recreation and its progress toward professionalism should review past editions of *Recreation Management*.

1961

NIRA established a Certificate Program. That program has served as the needed means of identifying industrial recreation administrators who have the qualifications entitling them to be considered as professionals. A Certified Industrial Recreation Administrator (CIRA) has passed a written examination and has (1) five years' experience as a recreation administrator or supervisor with a minimum of one year in employee recreation, or (2) has a baccalaureate degree in Industrial Recreation or a related field and one year's experience in employee recreation, or (3) a baccalaureate degree and three years' experience in employee recreation. Thus was established another of the criteria of professionalism.

1966

This year marked the conclusion of an extended, scholarly, and scientifically sound research project to discover the principles of industrial recreation in the United States. This was the doctoral dissertation of Donald E. Hawkins, Ed.D., New York University, 1967.[1]

Earlier and better known principles exist but they are of much less interest for several reasons. They are of less recent times. They are less complete. They were much less well researched and documented and, apparently, not validated at all. Of these, the two best known are: (1) G. Herbert Duggins and Floyd R. Eastwood: "Planning Industrial Recreation," published master's study, Purdue University, Lafeyette, Indiana, 1941; and (2) Chapter 4, "Recommended Principles and Policies," of Jackson M. Anderson's textbook.

The importance of validated principles for industrial recreation was well stated by Dr. Harry D. Edgren, formerly Professor of Recreation at Purdue University, in a speech to "professional industrial recreation educators" and entitled "Professional Assumptions and Alternatives":

> There are established principles (accepted practice based on valid experience) in every phase of the administrator's job. He does not function on the basis of tradition, trial and error, authority or current practice. He asks what are the accepted valid practices of good administration, of program, of facilities, of awards, or safety, of public relations, leadership and all other areas of the professional's job.[2]

1969

The Xerox Corporation industrial's recreation program, under the direction of William B. DeCarlo, CIRA, developed minimum requirements for industrial recreation staff standards. These requirements ranged from a degree in recreation and one year of experience to a master's degree and eight years' experience.[3]

This development was of great importance to the field of industrial recreation and its professionalization. Some other large programs have done likewise. When the practice becomes universal, we will know that full public recognition of the professionalism of Industrial Recreation Administration has arrived.

Although progress toward professionalism has probably been underway every year since its beginning, there was a notable and exciting splurge of renewed interest in professionalism in 1971 and 1972 as witnessed by events and the articles appearing in *Recreation Management.*

1971

Fittingly the renewed assault upon the objective of public recognition of the professionalism may have been inspired by an industrial leader. Robert W. Galvin, President of Motorola Corporation, made the keynote address at the NIRA national convention of 1971. In the course of his remarks, he made one especially prescient statement:

> In my estimation, industrial recreation is the newest *profession* [emphases added] in our industrial society. . . . Although organized recreation went back to the very founding of this country, it is somewhat ironical that it was only a few years ago that this type of formal activity became recognized in industrial society. It is not without reason, therefore, that maybe we've only written the first chapters of what you can accomplish for American industry.[4]

Robert W. Galvin is included in *Top Management Speaks* and a statement by him in that publication prefaces this chapter.

It is probably not coincidental that three of the leading industrial recreation professionals, all of whom were almost certainly present when Galvin made his speech, collaborated on an article of professionalism for the November-December, 1971, edition of *Recreation Management.* Among these excellent points made by that article, "Professionalism in the '70s," one dealt with the need for higher educational achievement:

> An extended education as a prerequisite for professionalism appears to head the list of desirable qualifications. Coupled with this education should be the enthusiasm and zeal necessary to continue to learn and practice the profession to the highest degree, thus culminating that which sets the professional apart from all others.[5]

In that same issue of *Recreation Management,* another article was "Professionalizing Industrial Recreation" by Ewen L. Bryden, Ph.D., CIRA, Professor of Recreation at Eastern Illinois University. It emphasized the need for:

1. stronger professional certification standards,
2. the development of program objectives, and
3. internships for students (pp. 22–23).

Still a third article on professionalization in that issue was written by Mel Byers, CIRA, long one of the prime movers for greater professionalization. In "Telescoping the Future" (pp. 25–26) he stressed the need for breath of academic preparation for the role of industrial recreation leader.

> The recreation administrator of the future must be well qualified in many areas. He will have to be familiar with psychology, promotion, journalism, personnel administration, recreation, employee services, accounting, selling, public speaking, marketing, labor relations, and business administration.

1972

Through the pages of *Recreation Management* we see the thrust toward professionalism continue uninterrupted in 1972. The January/February issue presents "Industrial Recreation: An Overview by an Industrial Psychologist" (pp. 16,21). John H. Rapparlie, Ph.D., an industrial psychologist for Owens-Illinois Corporation, reminds the readers of (1) one of the most basic of criteria for professionalism, and (2) the unique character of industrial recreation.

> The individual motivated to improve his managerial performance must also work toward the success of the business enterprise. . . . This means dedication towards contributing to solutions of problems encountered by that business or enterprise which involves knowledge as well as skills. It certainly includes adequate understanding of industrial economics and how the business system operates in our society.

In the March 1972 issue there is an interesting article titled "Industry and Education," by William E. Wolf, a student member of NIRA and a recreation major at George Williams College. Of special interest is his confession that from "my own experience as an undergraduate in recreation, and also from the general feeling that I received during interviews conducted with NIRA members and professional educators, I do not feel that the graduate in recreation has been prepared to handle the administrative aspects of a job in industry" [p. 10]. On the next page, he makes the astute observation that "today's campuses are almost entirely void of any knowledgeable information regarding industrial recreation." We would add that the situation has not notably improved today.

Wolf suggests that: "With the full cooperation of industry and education, they [NIRA] must research and publish a suggested curriculum background for a professional recreation manager."

It is evident that even as there was advocacy of changes for greater professionalism, change was taking place.

Evidence of Change

The NIRA Research foundation contracted for two research studies (1967 and 1971) concerning the characteristics of industrial recreation administrators. Both studies were done by Robert A. Frembling, a Recreation Professor at California State College at Hayward. A comparison of some of the more important findings appeared in the March 1972 issue of *Recreation Management,* (p. 60). These are shown below: Figures are all percentages.

	1967 Percent	1971 Percent
Have attended recreation workshops	27	53
Have attended recreation seminars	35	53
Attendance at NIRA conferences	35	100
College graduates	49.8	56.4

Other Indications of Change in 1972

In June 1972 NIRA Headquarters in Chicago acquired its first full-time student intern, Greg Demko of Western Illinois University. This practice has been continued since.

Recreation Management carried two more articles about professionalism, both by Dan L. Archibald, CIRA: August, "CIRA'S Obligation to the Profession," and, September, "Not Just a Piece of Paper." In the latter he recommends "even greater emphasis on the testing procedure in the NIRA certification program." [pp. 26–27].

Likewise indicative of the growing interest in professionalism is the item in *Recreation Management* [September 1972, p. 8] which notes industrial recreation developments in Great Britain. It reported that the Recreation Manager's Association of Great Britain "voted to allow students to become members of the association [and reflected on] the growing trend for more intensive training in recreation."

The Use of Student Interns

Between 1972 and 1978, the trend toward increasing concern about professionalism continued, largely supported by the same NIRA professionals mentioned in respect to 1971 and 1972, but with the inclusion of newer and younger professionals of energy and enthusiasm.

There probably was an increasing use of interns, but there is little information about that practice because it was not included in any of the NIRA surveys. We do know that by 1975 the Pratt & Whitney Aircraft Club, under the direction of Von Conterno, CIRA, had developed such a program, because an article in *Recreation Management* in April of that year was coauthored by a student intern ("What's In Store" by Von Conterno (CIRA) and Rich Dowdall, (pp. 36–37).

Also, beginning in 1975, Xerox, under the management of William B. DeCarlo, CIRA developed an ongoing plan with Springfield (Massachusetts) College for the use of student interns. This plan called for continuous contact and cooperation between the college (Dr. Donald Bridgeman, Director of Community and Outdoor Recreation) and the company. Most notable was the care with which the internship was planned and implemented, to the advantage of both the company and the students concerned. At the end of his nine-month work/study program, Intern Scott Baker working under the direct supervision of Xerox's David Baker, CIRA, (no relation) expressed the opinion that "you couldn't pick up half of what I've learned here in a classroom."[6]

Growing Interest in Professionalism

Although the interest and commitment to professionalism of the part of NIRA's leaders has been steady and unceasing, the expression of general interest in the matter seems to have come to the surface at intervals, increasingly closer together. Certainly this is true if we are to judge by the pages of the professional magazine, *Recreation Management*. We wish briefly to note four so related articles in March and April issues of 1977.

March. "Are We Professionals—or Batboys?" by Melvin C. Byers, CIRA, NIRA Consultant.

> Concerned with the need for public recognition of Industrial Recreation Administration as a profession, Byers asks the readers a number of questions, of which, for brevity, we select only a few. "Are we concerned with the manufacturing, products, and services of our company? . . . Do we recognize the unique role we are playing in business and industry or do we consider our position the same as municipal recreation administrators? Do we find it difficult to distinguish the difference?"

April. Three items, including the one about student intern programs already noted. (1).

(2) "Students on Your Staff: Valuable, Inexpensive Help," by Karen Bullock, a recent graduate of the University of Toledo, and then an intern at the *Toledo Industrial Recreation and Employee Services Council (TIRES)*, of which Mel Byers is the Director. A good article on the

value of student internship, an especially notable point was that, "with more exposure to business procedures and to actual problem-solving situations (as opposed to simulated in-class projects) the student gains better insight into the employee recreation field" (pp. 14–15)

(3) "A Look at Employee Recreation in England," a letter received by NIRA Secretary, Miles M. Carter, CIRA, from a member of the Recreation Managers Association of England. A story of what one progressive English company was doing in industrial recreation, it again signifies the professional interest in industrial recreation throughout the world.

The Growth of Local Industrial Recreation Councils

A few industrial recreation councils, local organizations of industrial recreation leaders and companies, have existed for several years, but the idea has been catching fire only in recent years. Nineteen are listed in the NIRA Directory for 1977 and new ones are being added yearly. The San Diego Industrial Recreation Council became a full member of NIRA on January 1, 1978. The special importance of this development to the growth of professionalism is that they provide "a new means of social contact for recreation directors and a medium through which they can exchange ideas."[7]

The growth of the local councils is strong evidence of the growth of professionalism in industrial recreation. In the article cited, Dick Brown, the author, adds:

> There are many of us on the NIRA Board of Directors who are excited about the future of the IRCs. We see them as natural extensions of NIRA and its services beyond the national and regional scope into meaningful, active, local organizations.

Although we, too, can look the development of IRCs with enthusiasm, we cannot be equally enthusiastic about the status of education for industrial recreation administration. To that topic we now turn our attention.

Education for Industrial Recreation

In recent years there has been a marked increase in the number of institutions of higher education offering academic programs in recreation and also in the number and variety and level of courses in Recreation. A survey of United States and Canadian Colleges and universities revealed that "as of March 1974, over 380 institutions have been identified as providing professional preparation in parks and recreation."[8] This may be compared to the 214 institutions offering such curriculums in 1969.

An Analysis of the Catalog

The curriculum catalog compiled periodically by the Society of Park and Recreation Educators is of interest to us for several reasons, the most important being that it is at this time the only source of information on nationwide United States and Canadian curriculums and courses of relevance to industrial recreation. It should be realized, however, that though many courses listed therein are of great relevance to industrial recreation, others are of little or no relevance, especially as compared to the value of other courses outside the field of general or parks recreation. The 1974 catalog is very well done and worthy of our attention in several respects.

Departments Offering Programs

The vast majority of programs are offered by a college of Health, Physical Education, and Recreation. Some are offered by a college of Education.

The Faculty

Most of the professors have advanced degrees in Physical Education or Recreation. Some have degrees from such diverse programs as Psychology, Law, Industrial Arts, Horticulture, Forestry, and Theology.

Advanced Degree Programs

The 1974 catalog listed some sixteen institutions which offered doctorates in some form of recreation and a great many more which offered work at the master's degree level. Leading states in terms of institutions offering advanced recreation degrees were California with 11, followed closely by New York with 9. Further down the list were Michigan, Illinois, Indiana, Maryland, Tennessee, and Texas.

Curriculum Options

In the Preface of the 1974 Catalog, an "option" is defined as "an area of specialization defined as a definitive area of study with several specialized courses completing a required core area of general park/recreation classes." Some of the more popular options listed were as follows.

Options	Number of Colleges Offering
Recreation Leadership	165
Municipal Recreation/Parks	146
Outdoor Recreation	120
Camping	96
Parks/Resource Management	70
Therapeutic/Ill/Handicapped	85
School Recreation	59
And far down the list:	
Industrial Recreation	22
(of these, 7 were two-year institutions)	

In studying the Catalog course listings, it was impossible to ascertain why any institution should have listed industrial recreation as an option because there were almost no courses specifically directed at industrial recreation. Possibly that option was listed because those schools offered their recreation majors the opportunity to intern in an industrial recreation program.

We do know that here and there specific courses for industrial recreation are taught, but we believe they are offered only occasionally and are usually taught on a part-time basis by professional industrial recreation administrators. Those professionals who do such teaching demonstrate their professionalism and dedication to industrial recreation inasmuch as the compensation for teaching on a part-time basis is far from commensurate with the effort and trouble necessary. Here again there is a paucity of information. We can identify only two such professionals (CIRAs William DeCarlo and Mel Byers), and we think that such an item might well be included on future NIRA surveys of its professionals.

Here and there are a handful of full-time professors fully qualified to teach specific industrial recreation courses, among them two who have backgrounds of work in the field: Ewen L. Bryden, Ph.D., CIRA, of Eastern Illinois University; and Donald E. Hawkins, D.Ed., of George Washington University. There may well be others unknown to us.

With the passage of time and the greater interaction of industrial recreation professionals with colleges and universities, especially through internship programs, the number of well qualified, full-time professors undoubtedly will increase. Meanwhile, there are courses offered by well-qualified faculty of general recreation programs which will well serve the future industrial recreation professional. Additionally there are a great many valuable and highly appropriate courses offered by well qualified professors of colleges of Arts and Sciences and Colleges of Business.

Recreation Courses

Actually the 1974 Catalog lists so many apparently relevant courses in recreation, that it is not feasible to list them all. Under one title or another, however, a fair sample includes: History and Philosophy of Recreation, Recreation in American Life, Recreation Leadership, Philosophic Foundations of Leisure, Leisure in Contemporary Society, Organization and Administration of Recreation, Recreation and the Community, Urban Recreation, Staff and Facility Management, and Research Methodology. Nevertheless, not even the total of all recreation courses offered by higher education will of themselves adequately prepare college graduates as professional recreation administrators. Therefore the academic preparation of future industrial recreation administrators remains of concern to the profession in general and NIRA in particular.

Certainly one of the reasons why there are not more courses and even complete programs for industrial recreation is that the institutions of higher education have not been made familiar with industrial recreation and do not know what is needed. The growing interaction of industrial recreation professionals and insititutions of higher education will bring about the necessary programs in time. Meantime, because the need is urgent and presently enrolled students need guidance, we do make a number of suggestions.

Our suggestions are made on the premise that a baccalaureate degree should be the minimum level of academic preparation for a professional industrial recreation administrator. We have adopted this premise not because a baccalaureate degree is historically the foundation of professionalism, but because we are convinced that is the minimum amount of academic preparation which can prepare a person for the difficult and complex role of industrial recreation administrator. Actually the needs cannot be fully met within a four-year program.

Two-year Programs

Within every profession there are a large number of duties and responsibilities that can be assigned to persons who do not have the extensive education, knowledge, and skills required of the professional. These nonprofessionals may be and often are extremely valuable specialists. Sometimes these specialists are called technicians; sometimes, paraprofessionals. However designated, these specialists generally are more interested in the "what" and the "how" than in the "why" of an occupation. In the field of industrial recreation, such a paraprofessional would be a valuable member of the staff but not the administrator. Basically, the two-year graduate would more often be engaged in face-to-face contact than with the leadership role. We could see the two-year graduate performing specific responsibilities under the general direction of a professional industrial recreation administrator.

In some instances, also, we can envision an especially capable two-year graduate assuming charge of a small company program, if serving rather immediately under the supervision of a knowledgeable director of personnel or like officer. However, before progressing to responsibility for a larger and necessarily more complex program, the two-year graduate would need to return for more extensive academic training or take the additional course work while in service. While acknowledging the skills and contributions possible by the two-year graduate, we would certainly reserve professional status and responsibility to those with greater academic preparation and more extensive knowledge and skills.

Academic Parameters of Four Year Programs

The definition of an industrial recreation administration program is difficult because of the unusual breath of preparation made necessary by the nature of industrial recreation. It will require the approximate equivalent of two-year's work to meet the general education requirements and complete the basic, building-block courses—though all these are not necessarily taken during the first two calendar years of college.

General Education

Whatever the program or major area of study, almost all colleges have some general education requirements which have to be met by all students. Every college should have. Whether or not called general education, there are courses which provide the student with those skills and the knowledge necessary, useful, and expected of every educated person. This core will include at least some courses in English (Communications), the social sciences, the natural and physical sciences, and the humanities.

Although directed essentially at the skills and knowledge useful to the laymen, most, if not all, the general education courses also contribute markedly to the occupational success of the graduate. Although English or Communication courses are general education requirements, proficiency in the use of the language is also a must for success in most professions and certainly in industrial recreation.

Foundation Courses

Foundation, or basic, or building-block courses are those courses which prepare the student for more advanced and technical or professional courses. Of these there are many. Examples may include some mathematics courses, an introduction to accounting, basic computer science, or survey courses of the area of specialization. There are a great many others, and their nature and number are determined by the student's program major.

Technical or Special Courses

These courses, most of which will be taken in the last two years, are those which are specifically related to the student's vocational skills and knowledge. For industrial recreation majors, the real pinch would develop here, for there is no way that in those two years the student could take all the courses which are needed by the industrial recreation director. The major in this field needs a strong grounding in all of the following areas, each of which might constitute the entire area of concentration for other majors: recreation per se, business management, personnel management, and social services.

It is particularly in the area of business management that the industrial recreation major has needs beyond those of the recreation major per se. There are three important reasons why the industrial recreation major needs a strong business background. (1) The larger and more sophisticated industrial recreation programs are big business. Annual budgets run into the hundreds of thousands of dollars; the buildings and other facilities are extensive; and personnel supervision is demanding. (2) The industrial recreation administrator must sell the recreation and services program to Management in respect to company goals. Business administrators value efficiency, organization, and capable management. Therefore the industrial recreation administrator should demonstrate those qualities. (3) The beginning industrial recreation administrator will almost certainly be required to assume some other company responsibility in addition to the work with the industrial recreation program. The academic program needs to be planned accordingly.

Among the more common nonrecreation assignments of industrial recreation administrators are: editor of plant newspaper, employment manager, assistant personnel director, public relations coordinator, supervisor of food services and/or vending operations, supervisory training director, grounds and park maintenance supervisor, employee communications and promotions manager, and employee services and benefits coordinator. The capability of handling these or other responsibilities may provide the needed entrance into the employment of a company and the opportunity to become established as an industrial recreation administrator. In any event, neither the skills nor the business experience will be wasted, for they can be put to good use by every industrial recreation director.

The Curriculum Framework

Within the parameters described, we make the following suggestions for a baccalaureate program in Industrial Recreation Administration. Detail is impossible because of the differing curriculum requirements, the various course titles, and the limitations of course offerings by the different colleges and universities. Merely for convenience of presentation, we will assume an institution operating on a semester basis with course offerings of three to five credit hours (averaging four credit hours) and requiring 120 to 130 semester hours for graduation. In such a circumstance, we might assume that 30 to 40 hours would be devoted to general education courses, 30 to 40 hours devoted to basic or building-block courses, and approximately 50 hours to specialized or technical courses.

In all three major areas, the courses suggested will exceed the number of hours which the student can afford to devote to the area. Therefore, subject to the special requirements of the institution chosen, the student needs to select those courses most appropriate in terms of personal interest and career objectives.

Curriculum Suggestions

Generalized Course Titles	Approximate Courses	Credit Hours
I General Education Courses		
English (Communications)	2	8
Humanities (Introduction to Fine Arts)	1	4
Sociology	2	8
History	2	8
Political Science	2	8
Economics	2	8
Mathematics	1	4
Psychology	2	8
Natural and/or Physical Sciences	2	8
Geography, Geology, Earth Science	1	4

Curriculum Suggestions—Continued

Generalized Course Titles	Approximate Courses	Credit Hours
I General Education Courses		
Physical Education	1	2
Personal Typing	1	1
II Basic, Building-block Courses		
Principles of Accounting	2	8
Computer Technology	2	8
Business Organization and Management	1	4
Business Law	1	4
Business Communications	1	4
Statistics	1	4
Introduction to Social Services	1	4
Community Resources and Services	1	4
Coaching or Administration of Team Sports	2	8
Two fields of fine arts (two courses each)	4	16
First Aid and Safety	1	4
Organized Communication	1	4
Public Speaking	1	4
Sociology of the Family	1	4
III Specialized, Technical Courses		
Introduction to Industrial Recreation	1	4
Industrial Recreation Leadership*	1	4
Industrial Recreation Supervision*	1	4
Industrial Recreation Programming and Scheduling*	1	4
Industrial Recreation Internship	1–3	8–12
IV Other Highly Desirable Technical Courses		
Recreational Facilities Planning and Maintenance	1	4
History and Philosophy of Recreation	1	4
Leisure in the Contemporary World	1	4
Investments in Banking	1	4
Personnel Supervision (Administration)	1	4
Accounting and Fiscal Controls	1	4
Budgeting and Financing	1	4
Public Relations	1	4
Principles of Selling	1	4
Principles of Advertising	1	4
Recreation Research Problem or Thesis	1	8–12

This by no means exhausts the number of courses which might be very desirable under some circumstances. We have not, for instance mentioned a foreign language, but some colleges require

*If not available, substitute nearest courses in Recreation.

one for graduation. Also, the industrial recreation administrator might find a foreign language very helpful in some instances, as Spanish in the Southwest or in Florida; or French in parts of Canada. Additionally, there are other Business courses, Recreation courses, or Social Service courses that might be necessary in some situations, and highly desirable in all.

What this adds up to, of course, is that the baccalaureate degree should be the absolute minimum academic preparation and that a masters degree is desirable. Beyond that, there are graduate level courses in research methodology and more specialized administrative, philosophical, and theoretical areas that could easily justify a doctorate. It is entirely conceivable that in time the terminal degree may become a requirement for the largest and most sophisticated operations and for those who aspire to teach Industrial Recreation Administration at the college level. At the present time, we are most concerned about the requirement for a baccalaureate degree in Industrial Recreation Administration and where such a degree might be obtained.

Institutional Restraints

Today relatively few colleges or universities permit the student the flexibility of course selection that would be necessary for building the specialized and personal type of program recommended. Perhaps with this publication, institutions which purport to offer a degree in Industrial Recreation Administration will develop such a program. Meanwhile, there are a few higher education institutions where the undesirable compartmentalization induced by departmental concerns and jealousy has been ended to the extent that the students are free to cross over between colleges (of a university) and departments to allow the student to create, with the advice of a program counselor, a customized academic program in line with the student's career objectives. The University College of the University of Toledo has that flexibility now. Perhaps there are, or soon will be, numerous others. The beauty of such flexible programs is that they do not require large enrollments in a special program to make the offering of it economically feasible. Meanwhile, the question sure to be asked by every prospective student in an Industrial Recreation Administration program is: What are the chances of employment?

Opportunities for Employment

In terms of numbers of positions to be open at a given time for graduates of appropriate programs, one would have to be a seer to know. There are hundreds of positions for industrial recreation professionals, but how many are or will be open in the near future is a difficult question.

Jobs with NIRA Members

Among the NIRA members only, there are many companies which have more than 10,000 employees each. These companies should have need for several professionally trained recreation leaders. Many of them, as we know of Eastman Kodak and Xerox, do employ several.

In past years, many if not most new positions in the industrial recreation programs were filled with persons already in the employ of the company, whether or not with any special expertise. It is still true in some instances that, even where the program is headed by a professional, new positions often go to persons already employed and who have served in some capacity, perhaps as a volunteer or in a lower staff position, in the company recreation program. With the growth of professionalism, however, more and more new positions are filled by educationally qualified candidates.

As noted earlier, almost ten years ago Xerox developed employment standards for recreational staff members with the minimum requirement being a degree in recreation and one year of experience. Other companies have followed that pattern, and more will do so in the future.

Within the past several years, a growing number of industrial leaders have recognized the need for professional leadership for their industrial recreation programs and have been unwilling to settle for anything less. For example, when Pratt & Whitney Aircraft Corporation needed to fill the top position in its large industrial recreation program, it conducted a national search for the right professional person. Having found and decided upon Von Conterno, then in California, it made him an offer so good that he accepted. As a condition of employment, the company moved him, his family, and all his household goods from California to the company site in East Hartford, Connecticut, unquestionably a very expensive proposition. The willingness to go to such trouble and expense indicates how important it was to Pratt & Whitney to employ a topnotch professional. More and more companies now appreciate the need for professionalism in industrial recreation.

In addition to the NIRA members which have huge numbers of employees, there are presently 50 NIRA Organization members with 5,000 to 10,000 employees; 120 with 1,000 to 5,000 employees; and 143 more member companies with 1,000 or fewer employees. Some of the latter group may not be able to afford a professional recreation program administrator, but many of them are and do.

As an example, the Flick-Reedy Corporation of Bensenville, Illinois, still has fewer than 1,000 employees but has employed a professional (Arthur Conrad, CIRA) for several years. Flick-Reedy has had one of the model industrial recreation programs and been a leader in the evolution of the field. Four times the Flick-Reedy program has won the Helms award for the outstanding program in its class. All this because Frank Flick, President of Flick-Reedy, has become convinced of the value of industrial recreation and is one of its most ardent supporters. Smallness of size certainly does not eliminate the possibility of professional leadership.

Industrial Recreation Councils

Recent years have seen the acceleration of the development of local industrial recreation councils (IRCs), and it is likely that they will come to have a significant impact upon employment opportunities within the field of industrial recreation. New ones are being formed almost every year. Some of these have volunteer or part-time administrators, but as they grow in membership, full-time administrators become almost essential.

NonNIRA Opportunities

Thus far we have considered NIRA members only. In addition we should mention that there are many companies with industrial recreation programs that have not yet affiliated with the national association. Further one should consider the fact that a student who follows the outlines of the suggested academic program, but leans toward the business-oriented courses, would be very employable by the vast and growing field of commercial recreation.

Employment Opportunities for Women

Although the larger number of industrial recreation administrators are men, there have been and are notably successful women in the field. The outstanding example is Martha Daniell, CIRA, who was NIRA President in 1970–71. No other woman has yet received such exceptional recognition, but there are many more who have enjoyed a large measure of success and are active in NIRA. The number of women professionals is growing. Among the list of 21 persons honored by being accepted into the ranks of Certified Industrial Recreation Administrators (CIRAs) in 1977, five (23.8%) were women.

Industrial recreation is a field wide open to both sexes. It is, indeed a profession which can benefit by a good representation of both sexes.

An Assessment of Opportunities

Obviously "potential" opportunities do not provide immediate employment; and if there were a sudden large surge of graduates of industrial recreation baccalaureate programs, there might not immediately be industrial recreation administrative jobs for all. Yet it has been demonstrated many times that in newly developing occupational fields it is the graduates of educational programs which stimulate new job openings. Sometimes it is upon the application of such a graduate that Management first gives serious thought to the development of a program.

Moreover, as one employer hears from another what a program graduate has accomplished, a decision is sometimes made to hire such a graduate in order to reap the same benefits. The documented growing professionalism of industrial recreation will also be responsible for generating new professional positions.

Certainly it would be folly to limit the number of enrolled students to the number of current or annually recurring job openings. The job openings will increase. Also, among the graduates of industrial recreation programs, some will undoubtedly wind up in another, probably related field. There are millions of persons whose careers developed in fields different from what they originally intended. Thousands of engineering graduates are not engineers, law school graduates are not lawyers, and teacher education graduates are not teachers. Yet these persons are employed and few of them regret their professional education. In most such instances, we find upon close examination, that it was the capabilities developed in their academic preparation for one career that provided them with the knowledge and skills upon which they built an unanticipated career.

Wise Academic Planning

The wise student remembers the old adage: Don't put all your eggs in one basket. Thus the wise student follows a program with a specific ultimate objective but with one or more alternative possibilities. If the student follows the suggested guideline, there will be a number of business-oriented courses taken, but there remains the option of taking still more business courses and so becoming more employable in a business or industry, though in a nonindustrial-recreation capacity. Once employed, the qualified graduate would find opportunities to work with the recreation program, or if there is none, to begin the volunteer development of such a program. Either avenue may lead the able graduate ultimately to employment as an industrial recreation administrator.

Different students have different preferences, of course, and some students might want to take strong minors in some other fields, as social services, school recreation, or parks or community recreation. We do believe that it is the better part of wisdom for industrial recreation majors—all students, for that matter—to develop at least temporary vocational alternatives.

Early Involvement

The interested student should get involved at the earliest opportunity with industrial recreation. A convenient path for such involvement is provided by NIRA—the opportunity to become a student member of NIRA. The cost is less than that of many college textbooks (reduced by 50% if the educational institution is a NIRA member) and confers many benefits. They include a subscription to *Recreation Management* and the opportunity to attend national and local professional meetings. Moreover, student members and their mailing addresses, are listed in the Annual Directory of the National Industrial Recreation Association. They may also receive some research reports and find assistance in terms of their intern program. Also NIRA has a developing employment service.

The internship, coming toward the last of the academic program, will provide interaction with professionals in the field, but the student should strive to make such contacts earlier on a voluntary and individual basis.

No one wants to see a tremendous, immediate influx of students into industrial recreation programs. A gradual growth is more desirable so that employment opportunities can expand with the number of graduates. What is needed at once is better academic programs so that the graduates will be better prepared candidates for industrial recreation employment.

Quite realistically, industrial recreation is a visible and growing field for employment. Those who have the basic interest should follow their inclinations, being astute in the development of their own academic program and making early contact with the industrial recreation field. Those who are interested will want to know something about the financial benefits, which is the next and last topic of discussion.

Salaries in Industrial Recreation

Those who are obsessed with the idea of getting rich would do well to avoid the field of industrial recreation. Yet those whose genuine interest is in industrial recreation need have no fears of living a life of poverty. As a fair generalization, we may say that the pay in industrial recreation is roughly on a par with most other *salaried* professions. It is rather difficult to be very specific because of the lack of adequate data. We do have the benefit of the "Industrial Recreation Report of 1977." This report, researched and analyzed by Abbott, Langer and Associates, and sponsored by NIRA and the NIRA Research and Educational Foundation, does provide us with more hard data than has been available in the past.

As we generalize from that report, however, you should keep in mind that it reflects *only salary*. Today a very large part of the total compensation of employment may depend to a very considerable extent upon fringe benefits. In some cases, these benefits are quite substantial, perhaps including life insurance, family health and hospital insurance, educational benefits for the entire family, reduced prices for company products and/or services, and, sometimes, even company cars or country club membership. Since there is no data at all about total compensation, we limit our discussion to actual salaries only.

All the following data and generalizations come from *Industrial Recreation Report, 1977,* Copyright by and purchasable from Abbott, Langer & Associates, P.O. Box 275, Park Forest, Illinois 60466.

Average annual salary increased directly from lower to higher levels of recreation administration. While the highest level recreation administrator had a median income of $18,100, the fourth-highest-level recreation administrator had a median income of $10,800. (p. 10)

Analysis of the annual salaries of recreation administrators showed the average salaries to increase fairly regularly for the highest-level of recreation administrator on the basis of the number of employees eligible to participate in the recreation program. The highest level recreation administrator in companies with under 100 employees eligible to participate had median salaries of $15,750, whereas those where 10,000 or more employees were eligible to participate had average incomes of $22,600. (p. 10)

The length of employment with the present employer was a significant factor in determining pay level. . . . Those with under one year of service with their present employer had median incomes of $10,920; those who had been with their present employer for thirty years or more had average incomes of $23,983. (p. 10)

The following data was taken from the same report, Table 2, Annual Salaries by Job Level, p. 13 (All *Recreation Administrators*)

Job Level	Median	Mean	Lowest Reported	Highest Reported
Highest level Administrator	$18,100	$19,172	$10,740	$30,650
Second level Administrator	$14,160	$14,911	$ 8,320	$25,773
Third Level Administrator	$13,000	$13,300	$ 9,528	$18,820
Fourth Level Administrator	$10,800	$11,049	$ 7,000	$15,920

The report, which goes into great detail, also revealed the following facts.

1. There is a fairly strong relationship between salary level and length of experience.

2. Salaries tend to vary by geographical regions, being highest in the northeastern states, next highest in mid-eastern states, and lowest in the southern states.

3. Excepting only holders of associate degrees (two-year graduates), the study found a direct relationship between salaries and the level of education.

The *Industrial Recreation Report* serves to reinforce our subjective conclusion that one should neither go into nor stay out of industrial recreation administration on the basis of salary alone. Let us conclude that for the person who has the interest, the commitment to study and work, and the necessary dedication for the profession, industrial recreation administration is an attractive choice and worthy of the strongest consideration.

With these observations we conclude this chapter. The reader of it and the previous chapters is already aware of the importance of the historical development of industrial recreation and at least somewhat aware of the importance of its national professional association. These will be further revealed in the next chapter, "The Roots and Status."

Notes

1. "The Formulation and Validation of Principles for Industrial Recreation in the United States." Published on demand by University Microfilms, Zerox University Microfilms, Ann Arbor, Michigan. Copyright by Donald E. Hawkins, 1967.

2. Although we have a copy of this excellent speech, we have not the means of identifying either the date or the occasion.

3. *R.M.*, November/December 1971, p. 24.

4. Quoted by Arthur L. Conrad, J.D., CIRA, in a speech at the International Recreation Congress, Brussels, Belgium, April 5–7, 1973.

5. Dan L. Archibald, Martha L. Deniell, William B. DeCarlo. All CIRA, pp. 19,24.

6. David Baker, CIRA, and Scott Baker. "Student Intern Programs: Good for Industry and Schools." *R.M.,* April 1977, pp. 16–17.

7. Dick Brown, CIRA. "How to Start an IRC in Your Community." *R.M.* April 1976, pp. 26–27.

8. *Recreation and Park Education Curriculum Catalog: Comprehensive Information about Parks and Recreation Curriculums in Colleges and Universities.* Published by the Society of Park and Recreation Educators, 1601 North Kent Street, Arlington, Virginia.

Battelle Columbus Laboratories mens basketball.

Battelle Columbus Laboratories mens fitness center.

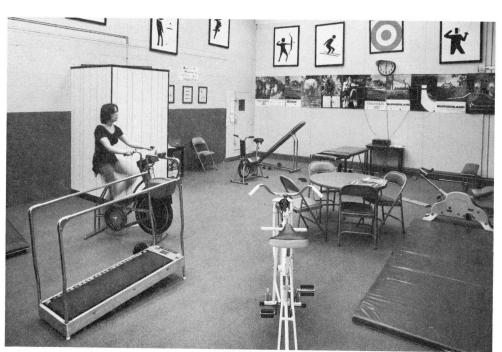

Battelle Columbus Laboratories womens fitness center.

10

The Roots and Status

TOP MANAGEMENT SPEAKS

Recreational activities have been an integral part of the Equitable way of life for our employees for more than 100 years.

The concept of employee recreation changes through the years. This is a natural development since recreation programs are initiated and administered by the employees whose attitudes and interests, in turn, are ever changing. The Equitable strives to keep its recreational programs apace with those changing times and tastes.

Despite these continuous changes, we believe it is essential to keep one factor constant: the recognition of the individual. This concern for our employees as individuals is natural for an insurance company which is traditionally committed to the personal security and well-being of people.

Varied recreational outlets for both young and older employees provide a unique unifying bond within The Equitable workplace. The satisfaction and recognition derived from participation in diverse recreational activities give a tremendous boost to personnel morale and can contribute importantly to an individual employee's own personal sense of development, particularly within the framework of his job growth. Recreation activities for and by employees also can create a genuine spirit of democratic participation that is to be welcomed and encouraged in a modern corporate environment.

John T. Fey

The Equitable sees its affiliation with the National Industrial Recreation Association as a means to improve the effectiveness of a balanced recreational program for its employees. And just as the company is committed to achieving continued financial security for its millions of policyowners and clients, so, too, the company is committed to maintaining the strength and stability of its own human resources, its employees.

John T. Fey
Chairman of the Board
The Equitable Life Assurance Society of
the United States

TOP MANAGEMENT SPEAKS

Even in these days of automation and the promise of increased leisure there is no real substitute for work, either for a company or for an individual. Hard work directed toward meaningful goals is still the only road to corporate success and personal fulfillment.

But the kind of work we most need today is not brute effort. It is the creative work that comes from disciplined and highly trained minds operating at peak levels of performance. Such minds do not operate in a vacuum. They are nourished by physical and mental well-being. They demand opportunities for enlargement, variety, refreshment, and recreation.

To invest in recreation is to invest in people. I know of no better investment a company can make than to invest in its human resources. Even more than its research programs, its facilities, or its financial strength, a company's human assets determine its future. We invest heavily in our human resources at Lockheed, and we give much consideration to recreation programs. They are a good investment.

Our company philosophy is to strive for excellence in everything we do, and recreation is no exception. We have allied ourselves with the National Industrial Recreation Association to improve the quality of our programs.

Daniel J. Haughton

We take a direct management interest in making them broad and creative and responsive to employee needs. We support them because they pay rich dividends in ways we cannot always measure but can never fail to see.

Daniel J. Haughton
Chairman of the Board (1976)
Lockheed Aircraft Corporation

We come now to the historical development of industrial recreation. We passed over it earlier as a matter of priority and because it fits better here. It is thus in close conjunction with the story of its important professional association, the National Industrial Recreation Association, and with the outlook for industrial recreation, which is the concluding chapter.

History is important. To some extent we can only know where we are—orient ourselves—by looking back upon the pathway by which we came. No important social development springs full blown into existence. Each has its faint, at the moment unrecognized, beginning. Through the lens of history, we can see the early and apparently unrelated phenomena become increasingly common, assume a faint outline, and gradually become the dominant pattern. Industrial recreation is an excellent example.

Social Forces

The development of industrial recreation can hardly be understood apart from the social forces of the 19th century: the conditions of labor, the development of labor unions, the organization of giant business monopolies, and the consequent titanic labor upheavals. Much of what transpired was a consequence of the so-called Industrial Revolution which resulted from the rapid change from human and animal power to mechanical power. That made possible the growth of the factory system which called for the massing of labor in industrial areas.

Labor

At the beginning of the 19th century, men, women, and children were working 14 hours per day in the winter time and from dawn to dusk in the summer time for wages which provided few luxuries and sometimes not even adequate necessities. The typical factory was poorly lighted, under heated, poorly ventilated, and had minimal sanitary provisions. The rights of labor, other than the receipt of a paycheck, were virtually nonexistent. The rather general attitude of employers was that they "owned" their workers during work hours and had good reason to control their lives at other hours.

These conditions were accepted by most workers, and change came slowly. Yet there was the beginning of trade unions among some skilled workers by 1820. This movement spread rather rapidly and there were trade unions in almost all of the larger industrial cities by 1835. With Jacksonian Democracy, there emerged a Workingmen's Party in Philadelphia in 1828. Organized to work for the rights of labor, its immediate objective was the 10-hour day. Others of its numerous goals were the abolition of imprisonment for debt and the achievement of universal, free, tax-supported, public education.

These concerted, voiced demands of labor were of some consequence. Free, tax-supported education spread widely during the second quarter of the century, though it did not become a reality in some states of the South until after the Civil War. The 10-hour day was also achieved, first in Philadelphia as the result of a series of strikes. President Van Buren decreed it for federal employees in 1840. New Hampshire established the 10-hour day by law in 1847, and a number of other states followed in rapid succession.

Ethnic Problems

In the course of the 19th century, there was the growth of class consciousness. Much of this appears to have been related to ethnic factors. In the forepart of the century, most of the immigrants came from Ireland or Germany. The majority of the German immigrants came with some financial means and were able to move inward and establish themselves on farms. The Irish were quite a different matter. After years of poverty and dissatisfaction, much of it related to absentee landlords, the Irish were already in poor straits even before the potato famines of 1846 and 1847. They came to America in droves and had little choice but to settle in cities and to work for whatever they could get. They lowered the wage levels for unskilled work. Moreover, being Catholic, they were disliked and distrusted by the majority of American Protestants.

Nevertheless, the Irish "problem" did not approximate the ethnic problems of the last quarter of the century when a rapidly increasing portion of immigrants came from eastern and southern Europe. These immigrants often knew little or no English and almost nothing of American culture. They were prone to hold to the ways of the old country. They did not mix well nor were they well accepted by the native American workers.

Monopolies

The Civil War Reconstruction Era saw the organization of gigantic American corporations with the power of monopoly and the disposition to make the most of it. These great corporations formed themselves into employers associations to deal with labor by means of the black list and by agreement upon wages to be paid. John D. Rockerfeller and Samuel Andres formed the Standard Oil Company in 1866. In the period of a few years, he and his associates established a virtual monopoly of the oil business as one after another of their competitors was eliminated. Other industrial giants were developing in other fields and, in time, most of the powerful corporations were acting in concert. Interlocking directorates were common, and a relatively small number of families or their representatives sat upon the boards of most of the great corporations of America. This was to the disadvantage of the public at large and to labor in particular.

Labor Unions

The Noble Order of the Knights of Labor was formed as a secret organization in 1869. Although nonviolent in principle, it found itself involved in a number of large and violent strikes. Although its secrecy was abolished in 1878, it had already become an object of public distrust and, after 1886, its membership and prestige declined rapidly. Its demise was hastened by the influence of new and vocal labor leaders from southern and eastern Europe who, contrary to the leadership of the Knights of Labor, deliberately encouraged class consciousness and class struggle.

By mid-century there were a few strong craft and trade unions which had national organizations. Among them were the Typographical Union, the Brotherhood of the Footboard (railroad engineers), the Iron Welders International Union, and the National Union of Machinists and Blacksmiths.

In 1881, Samuel Gompers and Adolph Strasser organized several craft unions into the loosely knit Trades and Labor Union of the United States and Canada. It was later reorganized into the more powerful American Federation of Labor, which grew stronger as the Knights of Labor declined.

Labor Troubles

The first great and violent strikes were against the railroads. The railroads had been badly hurt by the Panic of 1873 and by the unbridled competition among the lines as they sought to destroy each other and win a monopoly. By 1877, their financial weakness caused them to agree to end their competition and to divide the market and traffic between them. They also agreed to reduce the wages of all their employees by 10%. The employees of the Baltimore and Ohio Railroad had already had four wage reductions in seven years, and their reaction to the new reduction was strong and immediate.

On 16 July 1877, firemen abandoned their engines at Martinsburg, West Virginia. They then blocked the railroad tracks. That night, reinforced by thousands of other men who had had only irregular employment since 1873, they began using violence against railroad property. The rioting, burning, and plundering soon spread quite widely. Though much of the violence may have been the work of the mob rather than of the strikers, the strikers got the public blame. The strike, often accompanied by great violence, spread rapidly. Railroad traffic was tied up from the East Coast to the Mississippi River. There was incalculable loss of property and economic damage before federal troops and state militia were able to restore order and break the strike.

During the 1880s there was a short period of prosperity during which labor unions were able to gain more favorable working conditions, the 8-hour day, wage increases, and the recognition of the unions as bargaining agents. But hard times had returned by 1885 and most of the recently gained advantages were lost. Many employers took advantage of the competitive labor market to reinstate the 10-hour day, lower wages, and withdraw recognition of the unions. Violent strikes again became the order of the day.

Among immigrants from central and southern Europe were some social anarchists who formed what was called the Black International. Their avowed goal was the overthrow of capitalism and the establishment of a new social order. In May of 1886, the McMormack Harvester Works was on strike but was kept open by strike breakers. At a general strike meeting in Chicago's Haymarket Square, a dynamite bomb was thrown among police trying to disperse the meeting. One policeman was killed and many were gravely injured. The meeting had been addressed by members of the Black International. Eight of them were brought to trial, of whom seven were sentenced to death, though the sentences of two were later commuted to life imprisonment. There is much controversy about the justness of the convictions; but the public was much aroused.

Another particularly notable strike was that of the Amalgamated Association of Iron and Steel Workers against the Homestead, Pennsylvania, plant of the Carnegie Steel Company in 1889. The company built a high wire fence, hired several hundred armed guards, and operated with special strike breakers and nonunion workers. The violence was great and resentment was bitter. The strike ultimately was broken by the intervention of 8,000 National Guard troops who surrounded the factory. Union recognition was withdrawn by the company and those employees who were allowed to return to their jobs did so on company terms.

One other strike which calls for mention was against the George W. Pullman Company. Pullman had built a company town in a suburb of Chicago. The company owned the houses, the stores, the church buildings, the theater, and the park. Following the panic of 1893, Pullman laid off about one-third of the workers and cut the wages of the remaining by about 20%. Yet the company house rentals and prices at the company stores were kept at their former levels. The workers went on strike and proceeded to disconnect all Pullman cars from trains. They also succeeded in getting the support of the American Railway Union so that railroad freight and passenger service was at a standstill from Chicago to the Pacific coast. The strike was broken when President Cleveland, "in order to protect the U.S. Mail," had federal lawyers obtain injunctions and sent federal troops to enforce them.

Those strikes mentioned were only a tiny sample of those occurring. American historians believe that between 1883 and the end of the century there were more than 20,000 strikes and lockouts. It was a period of exceedingly great labor turmoil. The effect of the labor troubles and the other mentioned aspects of the nineteenth century greatly affected the development of industrial recreation, as should become clear in our next section.

Industrial Betterment and Welfare Work

The Industrial Revolution began in England and on the adjacent continent almost fifty years earlier than in America. Thus the great social upheavals and the need for the improvement of the lot of labor occurred there earlier also. Two instances of improvement programs interest us, as they are in some sense forerunners of industrial recreation though they sprang from two motivational extremes. In England, Robert Owen, well known as a philanthropist and socialist, built schools for the orphan paupers hired out to his company, provided good housing for his workers, and established a nonprofit company store. He operated from the motivation of philanthropy.

In Germany, Alfred Krupp of the great Krupp Iron Works did many of the same things. He built whole company towns besides providing his workers with insurance and pensions. However, Krupp's motive, fairly clearly, was simply to stem the rising tide of trade unionism. His was a *quid pro quo* arrangement (one thing for another) and his concern was for his own profits.

In the United States, as industrial recreation began, it was most often provided from one or the other of those two motives, though, obviously, there was some intermixture. Yet even those operating from philanthropic motives were inclined to be of a condescending nature and with a "holier than thou" attitude. Not until the end of the nineteenth century and the early years of the twentieth century do we see evidence of the mutuality theme—a joint enterprise of labor and capital, from which each could benefit. We might add that that theme is not universally accepted and implemented yet, though it is pronounced and strong.

Industrial Recreation Historian

By good fortune, there was a historian of the beginnings of what we now know as industrial recreation. He was William H. Tolman, Ph.D., and he devoted years of his life to the promotion of what he called "industrial betterment." He also collected all the information available about such matters and wrote the book *Social Engineering,* published by McGraw Publishing Company in 1909.[2]

Tolman, who was active and apparently pretty well known nationally and internationally, called himself a "social engineer." He described that "new profession" as persons "who can tell the employer how he may establish a desired point of contact between himself, his immediate staff, and the rank and file of his industrial army." [p. 3]

Industrialists had not earlier seen the need for another new profession which Tolman asserts he introduced in America upon returning from the Paris Exposition in 1900. This was the position of "social secretary," a person charged with taking a personal interest in the welfare of the workers and serving as a point of contact between them and management. At first, most of those positions were filled by women, but more men became involved as the position evolved into personnel departments and the position of personnel director.

Two other terms which came into common usage during this period were "industrial better-ment," and [industrial] "welfare work." Tolman created the former and, though he is somewhat ambivalent in describing it in other places, in his "Preface," which presumably he wrote last, he stated:

> Industrial betterment is more comprehensive, inasmuch as it concerns the efficiency of the worker as well as the plant, in other words, this betterment of the labor element is a cold business proposition and is undertaken commonly to get the best results out of labor. Welfare work concerns that improvement of the personal conditions of the laborer; this is often offensive to him and is resented by him, savoring as it does of paternalism and charity. [p. iii]

Social Engineering, not a literary masterpiece, is to a large extent, a collection of brief narratives of the actions taken by various companies in respect to industrial betterment or welfare work. Tolman made little effort to related the various industrial programs to underlying social and economic conditions. The threads or themes which run throughout the period—about 1850 to 1908—are those which, with the benefit of hindsight, we have been able to identify and have chosen to emphasize.

Motivation and Emphases

Tolman does force a revision of the often expressed view that industrial recreation began out of an interest in competitive sports and company sponsored amateur athletic teams. Such was not the case. In the latter part of the nineteenth century, a notable problem was the assimilation of central and southern European immigrants. There was much friction between the office workers, mostly native Americans, and the less well socialized and usually foreign-origin factory and warehouse workers. Because of the common illiteracy among the rank and file workers, manage-ment also felt it difficult to communicate with them. Also, because of the social conditions of the burgeoning cities, there was little after work entertainment for the workers other than the barrooms which flourished everywhere.

Consequently, in the beginning, the strongest efforts were directed at education—classes, libraries, reading clubs—and social activities designed to make workers into "ladies and gentle-men." Whether the desire or determination to improve the character and the lives of the workers was out of social service, in its most unpalatable sense, or was designed to increase production, the results were about the same. Both motives show through in a great many of the practices or programs related by Tolman. We quote three.

> 1. Mrs. Sidney Laughlin, mother, and George A. Laughlin, son, had heard so often that the saloon was the only club house for the working man that they were determined that this should not apply to the workmen of a company wherein the management was under their control. Accordingly they built a club house for the use of the employes of the Cleveland Axle Manufac-turing Company and the Canton Springs Company, at Cleveland, Ohio. The club is arranged to contain baths, a writing room, library or recreation room, billiard room, bowling alley, auditorium and game room, *but no cards* [emphases added. Gambling was considered another of the great vices of the working classes. [pp. 41–42.]

> 2. Club houses intended as a check to the drink habit so prevalent among the men by furnishing a place where intoxicants can be purchased only under certain well-defined regulations, and where various forms of wholesome amusement are provided to take the place of the demor-alizing features of the saloon, are provided by the Colorado Fuel and Iron Co. [p. 39]

> 3. One industrialist in Indiana tried to counteract the drinking habit among his men by a preaching service conducted by the different ministers in the city once a week in the factory. . . . This experiment was abandoned after 6 weeks as the owner "was doubtful of its success." [p. 42.]

Another example of the social reform motivation is revealed in the story of the Dodge Manufacturing Company which fitted up a dining room so that a banquet or dinner could be served. Later club rooms were added. In some respects the Dodge club was superior to many others, but we are particularly struck by Tolman's final paragraphs in its narration.

> The reading circle holds regular monthly meetings in the library, while stereoptican lectures are enjoyed biweekly. . . .
>
> It is the intention to make these quarters so persuasive in their attractions that they will be the means of bringing out all of the good, from a mechanical standpoint, that may be in one's make up. It is the sole desire and aim to place at the disposal of the employes that through which they can be bettered after factory hours. [p. 306]

Among those most dedicated to esthetic and educational improvement were the Roycrofters, of East Aurora, New York. Of them Tolman wrote:

> The central Roycroft building is devoted entirely to intellectual, aesthetic and social opportunities. . . .
>
> In the second story are rooms where evening classes are held, all open to the employes without expense save that each must purchase his own books. All the instructors are workers in the shop, and there are classes in modern languages, literature, history, and designing, with exercises in debating. A special feature is made of music, in charge of a musical director from outside. Instruction in voice and piano is offered employes free, and those who undertake such study are permitted to take a half-hour daily in working time for practice or lessons. [pp. 307–308]

Of all "betterment" facilities offered by industry in the nineteenth century, the most common were libraries and reading rooms. They may not have been first in the hearts of the employees, but they were most important in the values of the employer-owners. In the beginning of industrial recreation, the workers were seldom consulted about their preferences or desires. The employers most frequently made the decisions without input from the "beneficiaries." A good illustration of the conmon employer attitude in this respect is the following narration by Tolman, quoted in its entirety.

> In 1870 the Conant Thread Company of Rhode Island furnished the transportation for their employes to one of the shore resorts on Narragansett Bay. The employes were not paid, with the exception of those on salary, for the excursion day, nor were they consulted as to the nature or place of the excursion. Ever since the company has paid for the dinner on the day of the excursion. [p. 318]

Employee Association

Although, as noted, what we call industrial recreation began as company donated and dominated programs, it gradually became apparent to a few of the more discerning employers that better results could be obtained by giving the employees a large voice in decision making. A definite trend toward employee participation had been established, but was far from universal, by the end of the nineteenth century.

Credit for the first employee recreation association is often given to the Metropolitan Life Insurance Company in New York in 1894. Tolman does not mention such an association, but he does mention that the company established a lunchroom managed by the employees, which does imply the existence of an association. At least this is the period of time when employee associations began to appear. Tolman relates that the Ludlow Manufacturing Associates of Ludlow, Massachusetts, employed some 700 women in their mills. Many nationalities were represented. At some unspecified date, "recognizing a desire for self-government among the operatives and desiring to foster it, the management decided to establish an experimental girls' institute, which should afford useful instruction and opportunity for rational amusement." [p. 267] Cooking classes were so popular that three different sections were organized; and "Saturday afternoons a number of Polish girls met to learn how to make American dishes." [p. 268]

The Ludlow Manufacturing Associates also started a library in 1878, with one room set aside as a smoking and small game room. Yet, as Tolman recounts, "the attendance became so disorderly that after several forcible ejections the room was closed." [p. 312] However, provision for recreation facilities was made in a new mill completed in 1895, and an association was organized which proved successful. It was the "Men's Club" and had its own board of directors. It became predominantly an athletic club, though there were family nights; and the reading, social, and class rooms were also open to the families. The gymnasium and swimming pool were reserved for women and girls three days each week.

Other self-governing employee associations formed about this time were those of the Celluloid Company and the Steel Workers Club at Joliet, Illinois, both organized at least by 1899. The Electric Club was organized in 1902 by the employees of the Westinghouse Electric and Manufacturing Company and the Westinghouse Air Brake Company.

The club house of the Iron and Steel Workers must have been one of the most outstanding of its time. Tolman reports that it was built by the Joliet Steel Company at a cost of approximately $75,000, a tremendous sum at that time. The maintenance and expenses were paid by the company, and the building was leased to the Club for one dollar per year.

The Celluloid Club of Newark, New Jersey, was described at some length by Tolman as "typical" and a "kind of working model." There, separate clubs were organized by the employees themselves, "some for athletic games and exercises, others for intellectual training and improvement, and still others for mutual aid in cases of distress through sickness or death." [p. 298] Later, after forming themselves into one company-wide association, "the company gave them, at its own expense, a club house, furnished throughout with everything required for its various uses. Its cost was $40,000." [p. 299] The Celluloid Company paid the taxes and insurance on the club house and several times met "extraordinary expenses" but otherwise the association was self-supporting.

Community Associations and Activities

In the nineteenth century, a great many factories and their employees constituted almost the entirety of some communities. Thus it is not surprising that, as Tolman pointed out, "when an employer begins some form of industrial betterment for the individuals in his own employ, the scope of his work widens out so as to include the community. . . ." [p. 324]

This was eminently true of Peacedale, Rhode Island, dominated as it was by the Peacedale Manufacturing Company. That company, founded in 1801, was incorporated in 1848 and soon thereafter began efforts for "industrial betterment," which consisted of building company "tenements." Of these, Tolman wrote in 1907, "The company tenements are plain, well built, comfortable houses, and though not especially modern in design, are always kept in excellent repair." [p. 234] In speaking of the various community organizations, Tolman explained:

> [They] are not generally in the hands of the manufacturing company as such, but have in most cases started and to a great extent carried along by the owners of that property. . . . Most of the organizations are thus village rather than company matters, but at the same time the company, its owners and employes practically make up the village. [p. 324]

Something of the same kind of development must have occurred in Ludlow, Massachusetts, home of the Ludlow Manufacturing Associates. At the time of writing *Social Engineering,* Tolman mentioned that a social secretary originally employed by the Associates had since "been engaged by the Athletic and Recreation Society, an organization of the men and women of the village of Ludlow." He added that "the society is the home for cooking, dressmaking, millinery, laundry, classical culture, swimming and dancing classes; the dramatic, orchestral, choral, football and baseball clubs; it has a bowling alley and a pool room." It was incorporated in 1896, the first incorporation of a recreation association that we know of. Also, the "social secretary" was almost certainly working as a *full-time manager* of the association, also a first so far as we have any inkling. [p. 86]

Forerunner Activities

There are few present-day activities or interests of industrial recreation which did not find some expression during the latter half of the nineteenth century. Notably lacking is group travel, though group excursions were not uncommon. A rather common feature of the period was the association-operated lunch room which served regular meals to the workmen. In the main, this has since been abandoned as not sufficiently profitable.

Physical fitness certainly is not a recent development. Even those employee clubs which most strongly emphasized social and intellectual activities almost always had some kind of game rooms and athletic teams. An excellent example is the community operated Carnegie Library in Homestead, Pennsylvania. Though its strength was intellectual, educational, and esthetic activities, the library club was also a strong believer in physical fitness. This was emphasized by Tolman.

> According to the prospectus, it is the object of the physical department to furnish to the members the best possible opportunity for the development of sound, active, enduring bodies, thus enabling them to pursue their vocations to the best advantage.
>
> It seems unnecessary to state that it is impossible to keep in a healthy condition, much less a vigorous one, without taking sufficient bodily exercise.
>
> There is no more effective nor convenient way of securing this vigorous, driving force than by making good use of a well-appointed gymnasium during the winter months. [p. 352]

The Roycrofters Club, mentioned earlier for its strong dedication to esthetic and educational development, also recognized the physical needs of the workers. *Social Engineering* tell us:

> The value of physical exercise is constantly emphasized at the Roycroft Shop. Fifteen minutes per day, *on the company's time* [emphases added], are devoted to exercise, either gymnastics indoors, under the guidance of a physical director employed by the company or to walks outside for those who prefer, participation being entirely voluntary. [p. 71]

Apparently the Roycrofter management recognized even then that one of the best means of motivating the rank and file workers for physical fitness is the involvement of the leadership. As it is reported of the Roycrofters, "Tramps across the country are urged at other times, and are frequently led by Mr. Hubbard [the principal stockholder and manager] or the physical director." [p. 71]

The Solvay Process Company of Syracuse, New York, is mentioned as one of several companies which built a gymnasium "for the employes and their children . . . where gymnastic classes are held under a competent teacher." [p. 317]

There was also group discount purchasing. The cooperative purchasing committee of the Seattle Electric Company made arrangements with selected companies "by which members of the association are enabled to secure the lowest possible prices for clothing, groceries, and all immediate supplies, and this same committee endeavors in other ways to promote the economic and business welfare of the members." [p. 92] Interestingly, too, the Filene Cooperative Association, of Wm. Filene's Sons Company, secured "good places for employes to go for the summer holidays."[3]

There are also several examples of employee association publications during this early period. *The Review* was published monthly by the employees of the First National Bank of Chicago. [p. 34] Another was *Echo*, published by the Filene Cooperative Association. Of it, Tolman informs us: "The company is in no way responsible for the articles appearing in the columns, as it assumes no voice in the management of the paper, and does not see the publication until it is printed." [p. 36]

Mutual benefit associations of employees existed in considerable numbers. We suspect that there are earlier examples, but we know of one founded by the employees of the Plymouth Cordage Company on 27 June 1878. [p. 149]

The Solvay Mutual Benefit Society was organized in 1888. Tolman also describes the beginning of the Industrial Mutual Association (IMA), which still exists and is going strong. Tolman notes that it began as a mutual benefit association of employees of vehicle manufacturing associations of Flint, Michigan. Later it expanded into an industrial recreation association and became communitywide.

Does It Pay?

The last chapter of *Social Engineering* is "Does It Pay?" It is quite revealing. Tolman quotes at length from a "letter of a prominent industrialist" who was quite negative.

> We went into it quite extensively. Whether it was that our people were a class of foreigners who did not seem to appreciate the work that we attempted to do, or whether their appreciation was misunderstood by us, I do not know. . . . There seems to be some invisible line between the employes in our factory and the office employes. . . . As far as our employes outside the office are concerned, we have fully made up our mind that it will be a matter strictly of business. . . . In other words, we shall buy our labor as we buy our material, and we are thoroughly convinced in our own mind that those who sell us their labor will give us as little as they possibly can for what they sell us without regard to whether or not we attempt to go more than our half way on some of those things outside of those which money can buy. [p. 356]

On the other side of the issue, we found the letter of the unnamed representative of the Acme White Lead and Color Works particularly noteworthy.

> Self-interest has certainly to be considered, as but few men are in business, carrying the burden with its endless cares and worries, because they love it; but rather to provide a competence for themselves and those dependent on them . . . [but] a business policy that eliminates sentiment, we judge *not* a good policy. . . .
> There are things, precious things, that money can never buy. It can and does purchase perfunctory service, but it cannot and does not purchase loyalty and goodwill, without which, no service is worth the having.

> There are contributory causes for our business success, but they are invoked by a system of management which recognizes in every other man, a man and a brother. . . .
>
> Goodwill is a mighty factor in business life, but it must be sincere, for unless a man has that goodwill in his own heart, he will certainly fail to inspire it in others. . . . [pp. 364–65]

It is of interest to examine the opinions of Tolman himself after he had spent many years studying the field of labor conditions and labor relations of his time. His basic conclusions are clear:

> A decade ago when there was a decided impulse toward some form of improvement, it was undertaken not through altruism but through necessity. The awakened intelligence of workmen began to voice itself in expressions that something more than wages was due. . . . To allay this feeling of smothered discontent, the industrialist was forced into attempts at betterment; he felt this step was necessary to hold his labor. Is it any surprise that some failures followed those attempts? Others, actuated by humanitarian principles, proved successful. . . . *All these attempts founded on mutuality endure; it was teamwork by capital and labor.* [emphases added. p. 365]

In the last paragraph of *Social Engineering,* Tolman returns to the key concepts: "Industrial betterment: and "welfare work"; and he disavows both. His concluding thought is expressed in one sentence: "It now seems to me that neither of the phrases adequately connotes the cooperative spirit which should exist between capital and labor, a mutual relationship which is best expressed by the term 'Mutuality'." [p. 366]

The Twentieth Century

As compared to the latter part of the nineteenth century, labor conditions and labor-management relations have been much improved in this twentieth century. There have been periods of strife and strikes, but there has been much less bitterness and almost no emphasis upon class conflict. With few exceptions, labor has accepted the American economic system. Labor's efforts have been directed, not at destroying the system, but at securing a fairer share of the benefits of American productivity. In this goal, labor has met with considerable success though new laws, judicial decisions, the greater strength of labor as it became better organized, and by the support of a larger segment of the American public.

Yet labor progress has not been even. Characteristically, in moving toward its objectives, it has advanced three steps forward and taken two steps back. But forward it has moved! In war times and whenever the economy was flourishing, there has been a large demand for workers, and labor has prospered. It lost some, but never all, of the progress in times of recession or depression.

Twentieth Century Sources

For the twentieth century we have not one but several excellent sources. For the period 1900–1920, we have *The Human Factor in Industry,* by Lee I. Frankel and Alexander Fleisher, New York, The Macmillan Company, 1920. Covering the first half of the century, there is the chapter, "The Growth and Development of Industrial Recreation in the United States," in Dr. Jackson M. Anderson's excellent, original text, *Industrial Recreation: A Guide to Its Organization and Administration,* New York, McGraw-Hill Book Company, 1955. Through them and archives, we have a considerable number of surveys made by the U.S. Labor Statistics Bulletins, the National Industrial Conference Board, and, later, by the National Industrial Recreation Association. After 1958, there is a running account of developments reported by *Recreation Management.*

Progress of Industrial Recreation

The progress of industrial recreation has pretty much followed the same time table as labor progress. Industrial recreation has thrived most when employment expanded and has been constricted when business was less profitable and unemployment high. Yet, withal, as a great many surveys show, the trend in the number of companies participating and the number of employees involved has been always upward. Industrial recreation was particularly stimulated by World Wars I and II, for during both there were huge numbers of displaced persons, military and civilian, for whom recreation and services were needed.

Internal Changes

In many respects the most interesting feature of the history of industrial recreation in the twentieth century is the series of internal program changes. These include changes in the kinds of activities provided or emphasized, changes in the administration of programs, and changes in the sources of financial support.

Program Changes

For the most part, changes in programs or program emphases were evolutionary rather than revolutionary. We have found it more convenient to organize the changes by their nature rather than by time periods, for sharp distinction in time are not often possible. Yet some changes are typical of certain chronological periods.

Educational Programs

In general, the educational programs so strongly emphasized in the nineteenth century were deemphasized in the first half of this century. There are few references to such programs from 1900 until rather recent times. In 1920, Frankel and Fleischer dispose of educational programs in a single sentence: "The formation of groups of workers into clubs for educational purposes had been encouraged by a number of concerns.[4]

Social Programs

During the forepart of the century, social programs were given increased emphasis. In comparing Labor Bureau Statistical Surveys of 1916 and 1926, Jackson Anderson observed "a substantial gain in both recreational facilities and activities. . . . "[5] The social activities were undoubtedly stimulated by the social conditions prevailing in World War I. They were later quite adversely affected by the Great Depression. A summary statement of the findings of a 1934 research project for the National Industrial Conference Board was quoted by Anderson: "Social and recreational activities . . . were considerably retrenched during the depression. Over 50% of the social and dramatics programs were dropped. . . . [p. 62]

Although there was a considerable resurgence of interest in such programs during World War II, they have since been maintained at a rather uneven level. In terms of percentage of total industrial recreation programs, social programs have tended to decrease.

Musical Activities

Musical activities, principally bands and choruses, were strongly supported during the first quarter of the century. For the company bands, the companies usually contributed uniforms, instruments, music, and paid for an outside conductor. Sometimes, they even paid for practice on company time. This was on the assumption that the bands developed *esprit de corps* and were a valuable advertising medium. Choruses were supported in much the same fashion and for the same reasons. Yet, being quite expensive, the musical programs also were casualties of the Great Depression. The National Industrial Conference Board report of 1936 estimated that 42% of the musical organizations had been discontinued.[6] They never again became so prominent.

Athletic Organizations

Athletic teams were common in the latter part of the nineteenth century but were not dominant then. Apparently athletics grew to be of increasing importance during the twentieth century, though it is impossible to attach precise dates. The athletic programs suffered fewer and smaller cutbacks during depressions and recovered more quickly. By the mid thirties, athletics dwarfed all other industrial recreation programs.

In 1920, Frankel and Fleisher wrote:

> A possible danger of laying too much emphasis upon winning the championship is that men will be employed essentially because they are athletes, and a disorganization of the spirit and discipline throughout the plant is likely to result.[7]

Their fears were amply justified. From the end of World War I until at least the mid-1960s, employees were hired for their athletic prowess and paid in line with their value to the company "amateur" athletic team. During that period, and even later, some industrial recreation program directors also were selected, or promoted to their positions, on the basis of their athletic prowess and reknown.

"Big time" amateur athletics eventually were doomed by the development for which they paved the way: professional athletic teams. Several professional athletic teams grew directly out of company sponsored amateur teams. *Business Week,* 11 April 1964, cites two football teams examples: the Chicago Bears, from the A.E. Staley Manufacturing Company team, and the Green Bay Packers, from the Acme Packing Company team. Support for other amateur athletic teams was withdrawn for the reason that the growth of a large number of professional teams made the competition for skilled players so expensive that even the largest of companies found such advertising too expensive. An example is the demise of the internationally famous Phillips Oilers basketball team.

With the end of the "big time" amateur teams, athletics remained an important but usually not dominant aspect of industrial recreation programs. The emphasis in recent years has been on intracompany competition and the involvement of the largest possible number of employees, regardless of previous experience or expertise.

Physical Fitness

Physical fitness programs have enjoyed unbroken support from their beginning in the nineteenth century. What Frankel and Fleisher reported as "the growing realization of the benefits of physical fitness" led a sizable number of companies to build and equip elaborate gymnasiums and employ qualified instructors. As occupations became more sedentary, more persons became interested in physical fitness, and it has grown increasingly popular.

Control and Support

Program Control

It has always been difficult to ascertain who effectively controls industrial recreation programs. No one has yet been able to develop a questionnaire which will reliably and consistently elicit that kind of information. The answers supplied often depend upon who fills out the questionnaire. Yet of the trend, we can be certain. Even in the latter part of the nineteenth century, some employers were giving their employee recreation associations a large measure of self-control in respect to their own programs and activities. That trend has continued and been accelerated in this century. It seems that by 1920, about half the programs were largely administered by the employee associations. The Conference Board survey at mid century indicated that "in three-fifths of the companies, the employee recreation program was administered by the employees themselves."[8]

Most recent surveys by NIRA appear to indicate that employees exercise significant control in 75% to 80% of industrial recreation programs. It is probable also that most, though not all, of the company-administered programs are those of relatively small companies. There are few, if any, instances in which the employees are given practically no voice.

Financial Support

Well before the end of the nineteenth century, some employers discovered that the workers did not want handouts but preferred to provide at least some of the program expenses. The trend since has gradually moved toward less and less company support. The companies still provide half or more of the costs in a significant percent of employee association budgets. Yet large and well established programs have proved equal to the challenge of providing most of their own support if (1) company support is not withdrawn too suddenly, and (2) there is a full-time professional director.

Over the years, industrial recreation programs have thrived or declined with economic cycles. As recreation associations become more nearly self-supporting, they are much less affected by business cycles.

Concluding now the brief history of industrial recreation, we move to the history of the National Industrial Recreation Association. It has had a large and favorable impact upon industrial recreation since its founding.

The National Industrial Recreation Association

The National Industrial Recreation Association grew out of the National Recreation Association. For several years, the latter association had included an industrial recreation section in its national conferences; but the directors of industrial recreation programs felt that their needs were not being adequately served by the general association. The National Recreation Association was chiefly responsive to the needs of the majority of its membership, who were directors of government parks.

The industrial recreation directors felt that their needs were different in several important respects:

1. Their clientele were not the public on vacation weekends or trips but were regular and permanent employees of industries.

2. Their support came largely from industry, rather than from a government agency.
3. Their programs and scheduling, unlike those of public parks, were greatly influenced by industrial work hours and company locations.
4. They were themselves employed by industry and hence felt a need to contribute to the goals of the employing company.
5. The needs of the working man were different from those of the vacationing public.

During the National Recreation Association Annual meeting in Baltimore in 1939, seventeen of the more prominent industrial leaders decided to form a new organization. Though the creation of a new industrial recreation organization had been discussed for some time, the actual impetus for the formation of a separate organization appears to have been provided by manufacturers of sporting goods. Not to their discredit, they saw in the further development of industrial recreation a rich field for commercial exploitation.

According to Dr. Floyd Eastwood, then a Professor of Physical Education at Purdue University, he was visited in the spring of 1938 by a former Purdue graduate, then employed as a public relations representative for Brunswick-Balke-Collender Company, a sporting goods manufacturer of Chicago, "who desired a national survey of the extent of bowling in the industrial firms of the United States."[9]

During the fall of 1939 and the spring of 1940, Eastwood, who had been prominent in the Baltimore discussions, met with the president of McGregor-Goldsmith Sporting Goods Company and with a representative of the Brunswick-Balke-Collender Company. Together they promoted the idea of a separate industrial recreation organization.

Subsequently, Eastwood invited the members of the Baltimore Committee and other national industrial leaders to a meeting at Purdue University in March, 1940. According to the Eastwood manuscript, twelve persons accepted the invitation and "many others wrote indicating their interest and support of the formation of a national organization." At that meeting, a preliminary constitution was accepted and the following officers were elected:

President:	Dr. Floyd Eastwood, Purdue University
Vice-pres.:	Victor Vernon, Director of Personnel, American Airlines, New York City
Vice-pres.:	John Ernst, Director of Employee Relations, Consumer Power Co., Jackson, Michigan
Treasurer:	Ralph M. Voorhees, Director of Industrial Relations Brunswick-Balke-Collender, Chicago

The name selected was the Recreation Association for American Industry, and the temporary headquarters were established at Purdue. Unfortunately for history, the official list of those attending that important and far-reaching meeting has been lost. But among those attending, in addition to the elected officers, Eastwood listed: Lawrence Wittenberg, Director of Industrial Recreation, General Motors, Detroit, Michigan; Harold Mayfield, Industrial Relations Director, Owens-Illinois, Toledo, Ohio; and the unnamed Director of Industrial Recreation, Hughes Tool Company, Houston, Texas.

Subsequently, Brunswick-Balke-Collender Company granted $2,500 for a national survey of industrial recreation. This was undertaken by a graduate student at Purdue University, L.J. Diehl, and published in 1941 as *Industrial Recreation: Its Development and Present Status.*

Also through the mediation of Brunswick-Balke-Collender, the Athletic Institute of Chicago, an organization of sporting goods manufacturers, agreed to subsidize the Recreation Association for American Industry by supplying funds for an office, a full-time secretary, and traveling expenses. The first office of the Association was at Purdue University, but in 1941 it was moved to Chicago, where it has remained since.

Research for industrial recreation, sponsored by Dr. Eastwood, was centered at Purdue for the next several years. From 1942 to 1945, the Athletic Institute provided funds for the Association. For reasons not known to the writer, the name of the organization was changed to the Industrial Recreation Association on February 14, 1944.

Originally, the Industrial Recreation Association was largely an eastern and midwestern association. From 1941 to 1945, the Association sponsored conferences at:

Philadelphia, Pennsylvania	Boston, Massachusetts
Waterbury, Connecticut	Portland, Maine
Schnectady, New York	Columbus, Ohio
Manchester, New York	Minneapolis, Minnesota
Omaha, Nebraska	St. Louis, Missouri

and Auburn, Alabama.

The Far West became involved with the move of retired professor Eastwood to California. He recalls a meeting in 1948 in California which was attended by only four persons. What has since been called the first western conference was held in 1949, when Eastwood arranged a meeting at which ten delegates attended. That meeting was dominated by representatives of the fast-growing aircraft companies: Hughes, Northrop, North American, General Dynamics, and Lockheed. The Western Region became a part of the Industrial Recreation Association, which led to the change of name to the National Industrial Recreation Association, on 10 June 1949.

As may be evident, NIRA began as a management's assistance and guidance organization. Its membership consisted largely of personnel administrators and staffs of industry, educational institutions, and the personnel offices of government agencies. With the years, it has greatly expanded its interests, its membership, and its services.

As NIRA expanded its membership, it became too unwieldy to conduct all activities from the national office in Chicago, so regional divisions were created, of which there eventually became nine. Seven of these are in the United States, one is in Canada, and the ninth, called International, consists largely of institutions in Mexico, although there are also representatives from El Salvador and the Virgin Islands. Each of the nine regions is represented on the NIRA Board of Directors.

In addition to the national Annual Meeting of NIRA, there are regional conferences, clinics, and workshops. As the national organization has become more complex in its membership and services, the regional activities have become an essential element. A map of the different regions follows on the next page.

Associate Members

Among the most unusual features of NIRA is its inclusion of Associate Members, a practice which dates from its initial organization. Associate members are companies which have a commerical interest in affiliating with the National Industrial Recreation Association. Initially, these were largely sporting goods manufacturers; but now there are a large number and variety of such members. The range includes, but is not limited to, amusement parks and commercial attractions,

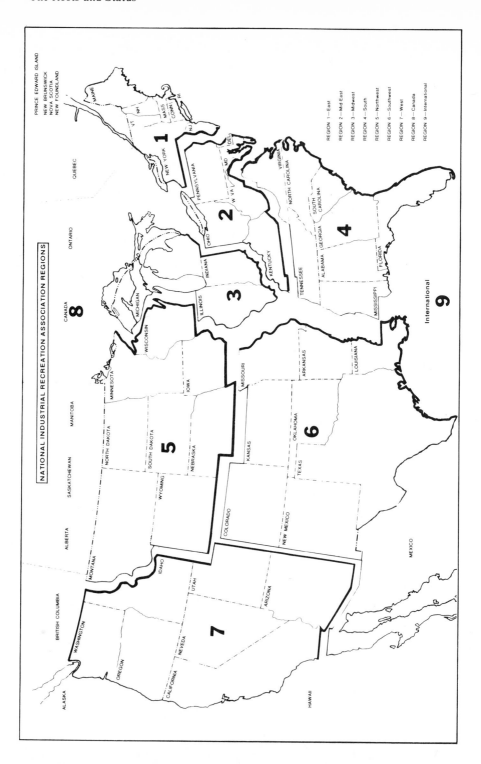

hotels and motels, airlines, steamship lines, bus lines, travel consultants and travel agencies. The associate members still include a large number of manufacturers of sporting goods equipment, but there are also large numbers of manufacturers and distributors of a wide range of products which have no connection to sports at all. Other associate members are motion picture producers and distributors, the tourist offices and travel bureaus of many foreign nations and of the various states of this nation, government departments of industrial development, city convention bureaus, restaurant chains, car rental companies, and publishers.

The associate members have open and frank relations with NIRA which are to the mutual interest of all parties. They provide considerable funding for NIRA, without which many important services would not be affordable. In return the associate members receive publicity, the opportunity to advertise and display their wares at the Annual NIRA Convention and at the many regional conferences and clinics. Many of the associate members have special contracts to provide goods or services to NIRA members at substantial discounts.

Initially the author viewed the existence of the commercial associate members with some suspicion. Upon closer examination, the conviction has grown that associate membership is a commendable, practical, and completely ethical arrangement. The openness of the relationship is in itself a safeguard against conflicts of interests. It has worked well from the very beginning of the association.

The associate members take a sincere interest in the affairs of NIRA and are allowed to elect one, but only one, member of the Board of Directors. Many of the associate members are competitors with each other, and we have never found an instance when the relationship of an associate member and a NIRA official or a recreation director has been subject to question. It is a mutually beneficial arrangement which we would recommend to other national organizations.

The associate members are important also at regional and local levels. Actually, one of the services that NIRA offers its members is the opportunity to participate in association-arranged contracts with associate members on very favorable terms.[10]

Developing Services

Through the years NIRA has continued to expand its services to its members. This has been possible through the dedication and ability of its presidents, the efficiency of its executive secretaries—of whom there have been only five since the beginning—and the loyal and generous support of the associate members. Much credit also belongs to its other officers and to those who have served so well on the Board of Directors. Beyond these, credit should also be given to those who have created and directed the growing number of regional councils and the local industrial recreation councils.

Membership

For some time the membership has hovered around one thousand companies and institutions, which represent hundreds of thousands of employees. The 1978 Annual Directory lists:

Organization and Allied Members, and Industrial Recreation Council Members	1,331
Associate Members	116
University and College Members	28
Student Members	196

Membership Benefits

Certainly one of the most important benefits is a subscription to *Recreation Management,* far and away the best source of information about developments and new ideas in the field of industrial recreation. Now a magazine of real professional quality, it is an invaluable asset to recreation program directors and to the Management of participating companies.

Of other benefits, there are so many, that it is inexpedient to go into any detail. We therefore content ourselves with listing the member benefits as they appear in the 1977 Membership Directory.

Other Benefits and Services

Program Manuals	Periodicals
Consultation Service	National and Regional Contests
Reference Library	Merchandise Discounts
Membership Directory	Certificate of Membership
Awards Program	Conferences and Workshops
Certification Program	Special Meetings
Post Conference Tours	Employment Services
Intern Program	Research Reports
Right to Hold Office	Right to Vote
Discounts from Associate Members	

We reemphasize the importance of *Recreation Management.* What is around the corner for tomorrow is already hinted at in its pages today. The successful innovations of one company today is translated into the trend of hundreds of companies for the future. And it is the future of industrial recreation with which we will be concerned in the next and last chapter of this book.

Notes

1. Rembert Patrick. et al., *The American People: A History.* Vol. II, 3rd ed. (Princeton, N.J.: D. Van Nostrand Company, 1962), p. 43.
2. There is another book: E.L. Shuey. *Factory People and Their Employers.* (New York: Lenilhon & Co., 1900), but it is much more limited in its scope and far less useful for our purposes.
3. "Employes Welfare Work." U.S. Bureau of Labor Statistics Bulletin No. 123, May 1913.
4. Lee K. Frankel and Alexander Fleischer. *The Human Factor in Industry.* (New York: The Macmillan Company, 1920), pp. 252–53.
5. Jackson M. Anderson. *Industrial Recreation: A Guide to Its Organization and Administration.* (New York: McGraw-Hill Book Co., Inc.), pp. 59–60.
6. Anderson. *Industrial Recreation.* p. 63.
7. Frankel and Fleischer. *The Human Factor.* p. 255.
8. Anderson. *Industrial Recreation.* p. 80.
9. Unpublished manuscript of Floyd Eastwood.
10. Richard M. Brown. CIRA. "How to Start an IRC in Your Community." *R.M.* April 1976, pp. 22–27.

11

Telescoping the Future

TOP MANAGEMENT SPEAKS

I believe we can do more to tap the great potential of industrial recreation. For years, industrial recreation has focused on the employee himself. Not it is being broadened to embrace the family, with more emphasis on family programming. The next big step? I believe industrial recreation must broaden its area of concern once again, and seriously embrace the whole community, from which the company must solicit its future employees.

A major development of our time is the discovery of the power of the private, commercial sector to help solve a wide variety of community problems. We see increasing evidence that business can hire and train the hardcore unemployed, educate functional illiterates, help reduce the rate of high school dropouts, help save alcoholics, even help rehabilitate criminals. Now I believe that industry should show what it can do to help meet community needs in the areas of recreation and health.

Industry today has vast resources in the form of physical recreation facilities, trained recreation professionals, and experienced recreation volunteers. I urge that these resources be put to work, more energetically and im-

Frank Flick

aginatively than ever, to serve not only employees but also the people in plant communities. To some extent, it already is being done, but more can and should be done in this area of industrial recreation.

Frank Flick
President
Flick-Reedy Corporation

TOP MANAGEMENT SPEAKS

Governments everywhere seek to improve the quality of life for their citizens. In Nova Scotia, where our government has made a major commitment to expand opportunities for individuals to participate in leisure activities of their choice, I look at the return from recreation budgets as one of our most outstanding benefits accruing from the wise expenditure of tax dollars.

Recreation dollars are definitely investment dollars! They return dividends to individuals, communities, business, and industry in the form of personal enrichment, improved health, higher morale, and greater productivity. Recreation dollars continue to work, improving the quality of life in our communities, long after they are invested. They bear interest in the form of human values as well as in long-lasting community resources.

We have seen tax dollars earmarked over the years in ever increasing amounts to combat crime and delinquency, to meet health and welfare needs, and to resolve the social problems of our day. But these are remedial dollars—expenditures made to resolve problems after they occur. As the alternative, recreation dollars offer a preventive medicine which provides a positive and human approach to resolving problems before they occur.

I see recreation as a two-sided coin: on one side sport, on the other culture. To get the best value from this coin, the individual must develop a balanced use of his leisure time, for his own benefit, as well as that of his community. Wise use of recreation by the individual is a stimulus to his life, and an inspiration to those around him.

Hon. A. Garnet Brown

In an age when unrest, unhappiness and dissatisfaction are prevalent, the wise use of leisure time holds out the answer to many of our problems—for all age groups, regardless of sex or racial origin, as well as the handicapped and disadvantaged.

The economic benefits of a dollar invested in recreational opportunities are estimated to return anywhere from three to ten dollars for every dollar expended. The social benefits to our people are immeasurable.

Hon. A. Garnet Brown
(1976) Minister of Recreation
Province of Nova Scotia

Without any psychic powers or gift of prophecy, it is yet possible to make some reasonable predictions for the future, at least for the near-term future. As students of society and of industrial recreation, we call upon our historical knowledge and our sociological understanding, mixed, perhaps, with just a "pinch" of artistic insight such as the poet uses.

Poetry aside, we all know that no event occurs in a vacuum and no event is spontaneous. Each event occurs within the context of an earlier event and, probably, the context of a future event. Historical events are like a long series of objects nesting within one another as, for instance, flower pots. Each fits not only the object before it but also the object which comes after it. It appears that by some inscrutable means future events work in conjunction with past events to shape present events.

It is interesting to note that the stream of social history is indeed much like a stream of water in respect to the forces that determine its contemporary characteristics. The character of a stream of water at any given point—its volume, speed, turbulence—is a product of both the terrain behind that point and of the terrain ahead. We all know how the mountain stream rushes and tortuously twists down the mountain side. Who among us has not also stood beside some clear and quiet stream meandering slowly along a broad valley and observed that, for no visible reason, it is beginning to pick up speed and change its character? Then we know that in that stream's future, somewhere beyond our line of vision, there is a sudden and sharp drop in the river bed, and the stream changes its character, picking up velocity and turbulence in anticipation of the coming waterfall. So moves and is affected the stream of social history—the story of society.

Nobody knows what society is or whether, in fact, it really exists except as a convenient symbolic concept. Yet, real or unreal, purposeful or not, each society has a character and peronality of its own—its own ethos. Though society itself can never be seen, its character can be seen, for its character is manifested in observable events.

Society, whether concept or reality, is more powerful than any of its constituent individuals or groups of individuals. Society's strength lies in the fact that it is able to find means to make its individuals *want* to behave in ways which are in the best interest of society. Yet society is in a constant process of change and, in its "wisdom," introduces ideas which at the time appear fruitless but which in reality are laying the groundwork for their acceptance at a later time when society is ready.

Our methodology is based upon the assumption that the future anticipates itself in the present, and that the present represents the transient conjunction of the past and the future. In a sense we feel that whatever there is in the present which is not fully explicable in terms of the past must represent the influence of the future and thus serve as an indication of things which are yet to come.

Thus we look to what has been and to what is in order to deduce what the future will be. For the most part, the conditions of the past and of the present have already been set forth in preceding chapters. Here, except for the concluding section on the international aspects of industrial recreation, we only emphasize the trends already noted and we take pains to note their relation to the future.

Dominant Long-term Trends

We observe that world societies, not merely the society of the Western World, have been moving, by fits and starts and sometimes almost imperceptibly, from the beginning of recorded history in the direction of greater democracy—political, social, and economic. During all that time, society has also been moving, albeit sporadically, toward greater concern for the individual—toward the individual's importance, freedom, welfare, health, opportunities, self-fulfillment, and happiness. To be sure, there have been the Genghis Khans and the Hitlers, but these have represented temporary aberrations and, as each has passed on, society has resumed its original direction.

That these views of the past and current trends are optimistic does not of itself deny their validity. They are based upon a fair knowledge and a rational assessment of the history of mankind and not upon wishful thinking. We realize that because the direction of society has been maintained since the beginning of history, that does not guarantee that the same direction will be maintained forever. Yet there is reason enough to suppose that the long-term direction will remain substantially unchanged for the near-term future, the next half century, at the very least.

These long-term trends are important to us, for industrial recreation is one of their products, and the future of industrial recreation is related to continued social change in the same direction.

The Infancy of Industrial Recreation

Frank Flick, president of the Flick-Reedy Corporation, an ardent and articulate advocate of industrial recreation, told the 30th Annual NIRA Conference a few years ago that "industrial recreation is still in its infancy."[1] The statement is almost as true today as it was then. Flick explained that "the full potential of industrial recreation is recognized by very few companies, . . . many small companies still believe that can't afford employee recreation programs, . . . many large companies have inadequate programs in relation to their size, . . ." and top management is not fully committed or involved with the programs that do exist. We would add that there are some large businesses and industries which have no company-sponsored recreation programs at all. As much as has been accomplished, there is much more still to do.

Democratization

The democratization of industrial recreation will continue for the reason that democracy is an accepted social goal. We have come a long way. In the 19th century, women were often ineligible for membership in the newly developing employee recreation associations. By the end of World War I, the bars had been let down for women, but there was still much social discrimination practiced. This was reported in a study by the Department of Labor reported in the 1919 Bureau of Statistics Bulletin, number 250. It reported that "in some of the mining communities the employers furnish club houses for their American employees but make no provision for the ordinary laborer, who is often a Mexican." One rather isolated company which felt the necessity of providing some kind of club house had three: one for Americans, one for foreigners, and one for Negroes. We can imagine that they were of quite different quality. Another not uncommon practice reported in that bulletin was the virtual exclusion of the blue-collar workers from the company clubs by the establishment of high club dues, "frequently $25 a year, a prohibitive amount for all but the better paid workers."

As shocking as those conditions seem to us, some not uncommon contemporary practices will seem as shocking to our posterity. There are still companies which have separate facilities and programs for the company "brass" and different and less desirable facilities and programs for the rank and file workers. Especially is this true in regard to physical fitness programs and facilities.

We wonder how long companies can continue to measure the value of human life and health almost solely by dollars and cents. How long can the life of an executive be considered as worth more than the lives of several "ordinary" human beings? How long will it be before it is finally recognized that there is no room in American business and industry for officers' clubs and enlisted men's clubs?

Even where there are no exclusive executive clubs or dining rooms, it still may be very difficult to persuade the "brass" to affiliate completely with the "ordinary" workers in full participation in employee recreation services and activities. Sometimes this lack of executive participation is from pure snobbishness. Sometimes it is from the belief that the boss cannot mingle with the workers socially without sacrificing authority and efficiency on the job. For those executives who are insecure and who may in fact already have been promoted beyond the level of their abilities, such fear is probably well grounded. Yet for those administrators of good character and the confidence that comes from great ability, closer acquaintance can only increase the respect for their authority on the job.

Fortunately change is occurring. Here and there, the "officers' clubs" are being discontinued, and the participation of executives in employee activities and services is increasing. The beneficent trend—for the companies and all concerned—will continue and the future will see still greater democratization of industrial recreation. Along with this change will come the more widespread acceptance of "mutuality."

Growing Mutuality

Industrial recreation is steadily moving away from the concept that industrial recreation clubs and programs are fringe benefits, that is, something the employers *give* to the employees in the hope of something in return—company loyalty, improved morale, and other hoped for benefits. A really successful industrial recreation program will never be manufactured at the bargaining table.

We learn but slowly! We return to two brief quotes from our chapter 10; the first by an administrator of the Acme White Lead and Color Works; the second, by Tolman.

> There are things, precious things, that money can never buy. It can and does purchase perfunctory service, but it cannot and does not purchase loyalty and good will, without which no service is worth the having.
>
>   . . . .
>
> It now seems to me that neither of the phrases [industrial betterment and welfare work] adequately connotes the cooperative spirit which should exist between capital and labor, a mutal relationship which is best expressed by the term "Mutuality."

Obviously the best of industrial recreation programs today are those undertaken with the full partnership of the company and the employee association, each secure in the belief that both mutually benefit from the employee activities and services. This approach is true of quite a number of companies today; but there are still disconcertingly large numbers of programs which are operated as fringe benefits, something *given* to the employees. Certainly the number of such programs and the extent of such thinking are declining now. The future will merely accelerate the trend toward mutuality.

Decision Making

Democratization will also affect the decision making process. Although all surveys tend to indicate that the employees, through their associations, are able to make virtually all their own decisions respecting industrial recreation, careful study and intimate acquaintance suggest that a very large number of companies still retain the authority to make the important decisions. This may be open or concealed. In many companies, designated administrators or the recreation program director—employed by and directly responsible to the company—reserve the right to veto any and all decisions of the employee association board of directors. In other instances, the association board of directors is intimidated by the participation of too many avowed representatives of the company on the board.

A healthier and more beneficial arrangement will be based upon mutual trust, so that rarely, if ever, would the company interfere with the operation of the employee recreation association. Some of this sense of trust and mutuality will come about quite naturally when larger numbers of administrators become full participating members of employee associations—not as administrators but as fellow employees.

Certainly neither we nor anyone else of whom we know is suggesting the complete independence of a company employee association. Rather it is a matter of respecting the rights and needs of each. Just as the company seeks to further the goals of the employee association, the association has an obligation to contribute to the goals of the company. It is sensible for the company to want the "good life" for its employees; it is sensible for the employees to want the company to be successful and earn a reasonable profit on the money invested and for the risk taken.

If there are sensitive areas in particular companies, where the control of the company is essential, these should be mutually agreed upon and the necessary procedures put into writing so as to avoid misunderstandings. Beyond such stipulated areas, the association and the company should operate on the basis of mutual trust. We believe this is a growing trend. Another trend relates to financial support.

Financial Support

There is a current trend, bound to be accelerated in the future, toward a greater degree of financial self-support by recreation associations. Though not all employers recognize the fact even today, the employees do not want handouts. Possible a larger number of companies will initiate industrial recreation programs when they realize that such programs do not necessarily derive all their funds from the finances of the company.

Probably in most circumstances it is desirable that the company build or provide financial assistance for the facilities of the employee association. Also, it is common and sensible for the company to underwrite the salary of the director and any professional staff. Beyond that, many employee associations are demonstrating that they can generate much of the necessary funds for the recreation program. As noted earlier, even some of the finest of employee association facilities have been built by the employees themselves, although some help was almost always given by the company. In order for an employee association to achieve so much financial responsibility, however, it must have the benefit of the leadership of a full-time, professional program director. Very little financial responsibility is likely to be achieved under nonprofessional, inexperienced, and part-time leadership. Finally, in finances, the idea of mutuality must always be kept in mind. The company and the employees together benefit from a good employee recreation association. Both the company and the employees should be prepared to pay part of the costs.

Nevertheless, as we see the growth in professionalism and in the numbers of professionally prepared administrators, we see the continued increase in the amount of employee association self-support. This is fortuitous for it will free industrial recreation to an unprecedented extent from the periodic set-backs occasioned by depressions or adverse business conditions. Largely self-supporting employee associations will be affected by depressions, but not to the extent of programs which are purely company financed. Thus the growth of industrial recreation should become more steady in the future.

Expansion

At one time, the bulk of the industrial recreation programs were in the northeast and north central states of the United States since those were the location of the major industries. Industrial recreation spread like wildfire in California as the new aircraft and related industries chose to develop there. Now we are witnessing the growing industralization of the South. We have seen an increase in industrial recreation there, but the peak is yet to come.

Likewise, in Canada, where large-scale industrialization was slower in developing, so was industrial recreation. But in recent years, Canada has expanded its industrial production greatly and, with that expansion, its participation in industrial recreation. Robert Wanzel reported in his doctoral dissertation that "the percentage of companies having recreational activities as a fringe benefit increased from 54.8% in 1967 to 61.7% in 1969."[2]

There is good reason to believe that the trend has continued and perhaps increased since that time. Also, as elsewhere, the idea of industrial recreation as a fringe benefit is giving way in Canada also to the idea of mutuality. There was, and perhaps still is, more of a tendency toward an unbalanced emphasis upon physical sports and physical fitness in Canadian programs. Yet there are Canadian programs, such as the well-rounded program directed by the veteran professional Murray Dick at the Dominion Foundries and Steel, Ltd., which provide examples for the desired breadth and diversification.

In Latin America industrial recreation is not yet well established but can be expected to grow with the increase of capitalization and industrialization. There is one particularly good example in Mexico which may contribute to the growth of industrial recreation there. That is the program of the ALPHA Group in Monterrey. The ALPHA Industrial Group Employees Association, headed by Jose Amores, seeks the total personal enrichment of all its members. It provides cultural, economic, educational, health, legal, and recreational support for the employees and for their families.[3] Latin America is a rich field for further development.

Local Industrial Recreation Councils

The increase in the number of local industrial recreation councils is an important reason to predict growth in the number of companies offering and employees participating in industrial recreation programs. The local councils make it possible for small companies to offer varied programs. They serve as local centers to educate business and industrial leaders to the benefits of industrial recreation. Especially are the local councils significant for their services in expanding industrial recreation from the traditional supporting companies to new kinds of operations—to banks, hospitals, retail stores, small service companies, and, in fact, to all manner of enterprises which employ people. The future will see a multiplication of local industrial recreation councils.

A Shorter Work Week

The growth of industrial recreation is predictable because of the trend toward a shorter work week and the emphasis upon the "good life" to vastly greater numbers of persons. It took more than a century to cut the dawn-to-dusk, six-day workweek down to the eight-hour day and five-day week that has prevailed for the last three decades. Now there is a growing crescendo of demand by workers, supported by public sentiment, to provide still larger blocks of time for workers to find recreation and personal enrichment.

One means of providing that larger block of leisure time was the introduction of the ten-hour day and four-day workweek. About 10,000 firms in the United States are operating on such schedules now. Another approach was the development of individual and flexible work hours, though the total time per workweek remained at 40 hours. Perhaps a million or more workers in the United States are working under the flexible scheduling concept. The *Toledo Blade* reported 15 May 1978, that the U.S. House of Representatives approved an expanded program of flexible hours for most federal employees.

Now the finger of fate points toward an actual reduction of hours in the workweek. In the United States the total number of hours actually worked by office employees declined about two hours in 1977 alone. Moreover, the United Auto Workers have begun a campaign leading toward a workweek of four, eight-hour days. In 1976, the UAW negotiated a new holiday program which provides "paid personal holidays" in addition to all regular holidays and vacations. Beginning in September of 1976, auto workers with at least one year's seniority received five personal holidays per year. The number of such holidays rose to seven in 1978. The goal of the UAW is to increase such personal holidays "until it achieves a four-day week."[4]

Also, the Delphi Forecast, based on surveys and studies of industrial experts and conducted jointly by the University of Michigan and the Society of Manufacturing Engineers, predicts that by 1990, the 32-hour workweek "will become the new standard for unionized industries."[5] The effect upon the popularity of industrial recreation is obvious.

Programming

As industrial recreation broadens its activities and its services—as the foremost associations are showing the way—it will appeal to an ever larger percentage of employees. As this happens, more and more employers will be moved to encourage and support industrial recreation. In his address to the 30th Annual Conference of NIRA, Frank Flick said (and the emphases are his) "I believe that industrial recreation can and should be *more purposeful, more significant, and should seek to achieve a wider range of objectives.*"[6]

As it does fulfill that need, and it is moving in that direction, it will certainly increase its believers, its supporters, and its membership. Industrial recreation is being more meaningful only through the leadership of its full-time professionals. We have here a beneficient circle: the increase of professional leadership will lead to the improvement of industrial recreation programs and services; and that improvement will, in turn, attract more participants and increase the demand for professional leadership.

Family Programming

In the quotation which prefaces this chapter, Frank Flick noted quite accurately that "for years, industrial recreation has focused on the employee himself. Now it is being broadened to embrace the family, with more emphasis on family programming." This is a change in process and not a goal fully achieved. There are still a distressing number of programs which are too narrow in their range of activities and have too little regard for family needs. It is important that there be activities in which families can participate as a family. Examples are family outings, family garden plots, family camp grounds, and organized tours for recreational vehicles.

To a greater extent then may be realized, the importance of the family is being reevaluated. Perhaps the great incidence of marriage failure has served to heighten the interest of persons in making their own marriages work, and less is being taken for granted. This may be part of the reason why the rapidly increasing divorce rate is beginning to level off. Although the divorce rate was 2% higher in 1977 than in 1976, this was the smallest increase for several years. Indeed it is quite a contrast to the 11% annual increases which prevailed from 1967 to 1973.[7] The declining rate of increase may foreshadow an actual decrease in the divorce rate in future years.

Also indicating the need for more family programming are the facts that it is the middle classes to which the bulk of employee association members belong, and it is the middle classes which are most family-oriented and which place the highest value on family activities.

Family Services

We see, also, an increase in family services provided by the employee associations. In some respects, this may be little more than the leading associations are providing now; but more associations will be involved. Among present services which benefit families are the ticket and discount merchandise services, which are quite common. Less common, but increasing, are association-operated stores which allow women or men to do part of their family shopping on lunch hours and without leaving the premises, a great convenience and money-saving also.

It seems most likely that the near future will see the introduction of one new and very important family service: day-care centers for the children of employees. So far as we have been able to discover, this service is not being provided on a regular basis by any employee recreation association now; yet it is a "natural" for employee associations. It is really curious that the idea has not already taken hold. More than a half century ago, the New Jersey Zinc Company at Palmerton, New Jersey, provided "a kindergarten at which those women who work and who have no one to leave their children can bring their little ones for a day's stay."[8] Industry in a few cases has followed that example, but not industrial recreation, for which it is even more appropriate. There are growing numbers of privately operated, commercial day-care centers; but there is still good reason for employee associations to begin their own day-care centers.

Association operated, nonprofit day-care centers, once established, can easily be self-supporting while simultaneously saving money for the parents. It is all the more feasible now that working parents receive an income tax break for money spent for the care of minor children. All that is necessary is for the employee association to do careful advance planning, research state and federal regulations, and ascertain and observe local zoning regulations and fire codes.

Organizing a day-care center is not a job for volunteers, but it should be well within the abilities and skills expected of the well educated and qualified professional industrial recreation administrator. That administrator will not operate the center, of course, but will need to stimulate and participate in the original decision making and preliminary planning. It will be the responsibility of the recreation director to make the necessary governmental contacts, arrange the business contracts, and look to the employment of a capable and qualified operator. Thereafter, the director will maintain only a supervisory relationship to the center as with other association operations. The suggestion for day-care centers does point again to the need for the continued professionalization of industrial recreation administrators, and it does have implications for the academic preparation of future administrators.

In operating day-care centers, employee associations will find they cannot avoid closer relationships with government agencies at all levels. Yet this closer relationship is inevitable anyway.

Industrial Recreation and Government

We agree with Frank Flick that, after increased attention to family programming, the next step will be for industrial recreation to "broaden its area of concern once again, and seriously embrace the whole community." This will not be a radical departure, for a number of employee association programs have been moving in the direction of greater community involvement for some time. It is only that the future will see a great many employee programs taking that pathway.

At the present time, the majority of industrial recreation programs closely involved with their communities are those located in relatively small communities. Yet there are examples of good community involvement on the part of large programs in sizable communities. A good example of close community involvement is the Industrial Mutual Association (IMA) of Flint, Michigan. The IMA belongs to the Flint community recreation association, which coordinates the use of all recreational facilities and through which all recreation events are scheduled. The IMA built a fine lodge which it leases for a token fee to the community Big Brother Association. The IMA also operates a community veterans counseling service. The IMA provides educational counseling service also and adds to that financial scholarships to nearby institutions of higher education.

Government Agencies

Once largely limited to private industry, industrial recreation is moving into closer relationships with government agencies. Many governmental units, as Metropolitan Dade County Florida, have large and flourishing industrial recreation associations for their employees and are members of NIRA. The employees at many U.S. Armed Forces sites have organized on their own and are members of NIRA also. Another sign of the times is the decision of the League of Federal Recreation Associations to join NIRA in 1977 as a "full membership Industrial Recreation Council." The LFRA serves a membership of more than 80 organizations which represent "over 300,000 federal servants."[9]

The Nova Scotia Conference

One of the most significant developments respecting industrial recreation in the past decade was the 1977 government-sponsored conference of government and industry leaders on industrial recreation held in the provincial capital of Halifax. The significant fact is that the impetus for the conference came from the government. Conceived and sponsored by the Nova Scotia Department of Recreation and the Department of Development, it brought together more than sixty important business, industry, and government agency leaders, all for the purpose of promoting industrial recreation. Four provincial governments other than Nova Scotia were also represented at the conference; and resource material generated by the conference was sent to all Canadian provinces.

That such a conference should be held in Nova Scotia came as no surprise. The Honorable A. Garnet Brown, Minister of Recreation, Province of Nova Scotia (1977), observed in his statement which prefaces this chapter:

> I look at the return from recreation budgets as one of our most outstanding benefits accruing from the wise expenditure of tax dollars.
> We have seen tax dollars earmarked over the years in ever increasing amounts to combat crime and delinquency, to meet the health and welfare needs, and to resolve the social problems

of our day. But these are remedial dollars—expenditures to resolve problems after they occur. As the alternative, recreation dollars offer a preventive medicine which provides a positive and human approach to resolving problems before they occur.

The Nova Scotia conference, entitled "Employee Recreation—An Investment in Your Company's Human Resources," marked the strong and outspoken commitment of the provincial government in the development of industrial recreation. Well planned, the conference secured the services of such outstanding industrial recreation leaders as Murray Dick, CIRA, Manager of Employee Recreation at Dominion Foundries and Steel, Ltd., and William DeCarlo, CIRA, Manager of the large Zerox recreation program at Rochester and a past president of NIRA.

At that conference, the provincial government officials went down hard and strong for industrial recreation. The conference summation was made by Nelson Ellsworth, the Recreation Department's Coordinator. He "indicated that the provincial Department of Recreation is prepared, through regional offices and other staff resources, to help Nova Scotia business and industry develop employee recreation."[10]

It is hard to discuss the Nova Scotia Conference without the use of superlatives. It signaled a new day for industrial recreation when it is not left to a few industrial leaders and to the professionals within the field to speak out for the social significance of industrial recreation. Whatever the initial, observable results of the Nova Scotia conference, the ultimate consequences should be tremendous.

We would suggest that representatives of institutions of higher education be invited to participate in any similar conferences in the future. It would also be extremely helpful if any provincial or state government would encourage its official institution(s) of higher education to provide professional, technical education in industrial recreation. That encouragement might take the way of offering some financial assistance.

The commitment of government and industrial leaders to industrial recreation can result in a magnification and enlargement of industrial recreation programs; but, in order to allow those programs to deliver the bright promises suggested, there must be an adequate and continuing supply of professional administrators.

Avenues to Professionalism

The growth of professionalism is virtually assured. A great and sufficient inducement for some companies to insist upon employing professional industrial recreation directors is the fact that only under such leadership are employee associations likely to become able to provide a reasonably large part of their own support. These companies will find it cheaper in the long run to pay the price of hiring professional administrators. Other companies, more fully persuaded of the value of industrial recreation to the companies themselves and to their employees, will want and insist upon professional administrators in order to achieve a program that will be the most productive of their goals and objectives.

As the market for professional leadership grows, so must the standards for professional recognition. Indeed, as the standards for professionalism are raised, the demand for professional leaders will increase. The two developments will go hand in hand. One problem, however, is the status of academic programs.

Academic Programs

At this time there appear to be no baccalaureate academic programs designed especially for the education of industrial recreation professionals. That is, there are no baccalaureate programs which contain even a few specialized, technical courses designed specifically for industrial recreation. Certainly programs in recreation are far from adequate for industrial recreation administrators. Programs for physical education majors are no better for our purposes. Industrial recreation is quite different from both, especially in its demands for business and administrative skills and sociological training.

Though special programs are needed, they will not come in one fell swoop. Institutions of higher education work with budget limitations, and they must see the prospect of sufficient numbers of students to make a program financially feasible before they can undertake to develop and offer it. On the other hand, the number of students will increase as specialized programs become available. Herein lies a problem, but it will be resolved. There is a special NIRA committee preparing to work with a few selected universities toward the development of such a program.

At the beginning, any such program may need to rely largely upon presently offered courses, though at least some courses might be tailored somewhat for industrial recreation majors. Two special courses can be required now however. One of them is Introduction to Industrial Recreation, based upon this textbook; the other is Internship in Industrial Recreation. Other courses can be developed one at a time.

To encourage registration of students in such programs, NIRA will give their graduates special recognition in the professional certification program. Most colleges and universities tend to be more flexible in their programs than used to be the case. Many institutions would now permit other education programs to be tailored, at least to some degree, to fit the needs of students who desire to become industrial recreation administrators. Among existing programs which might be suitably tailored are: Business Management, Personnel Administration, Public Relations, Journalism, Recreation, Parks Administration, and Physical Education.

The CIRA Program

The growth of professionalism and the demands upon industrial recreation administrators will almost certainly lead to more stringent certification requirements. The special NIRA committee is looking in that direction now. One development will be the ultimate requirement for a baccalaureate degree before the certification can be taken. We predict that NIRA will announce some date, perhaps five years down the road, to avoid unfairness to those who believe they are already started toward certification, when the only avenue to certification will be a baccalaureate degree.

With the changes in academic requirements for certification will come changes in the certification examination. The NIRA committee is now looking toward the substitution of objective criteria for examination instead of the present highly subjective and unsupervised essay examinations. NIRA is planning to achieve this goal by working with a national professional examination organization for the preparation and supervision of such examinations. NIRA professionals will provide the testing service with the objectives to be tested, and the testing service will devise the questions to test those objectives.

Another change which NIRA may some day consider is the addition of a requirement which exists now in many of the most highly professionalized fields: that is, a period, perhaps a year, of full-time experience working as a paid professional assistant under the personal supervision of a Certified Industrial Administrator as a qualification for full professional certification. That requirement would guarantee that a candidate for certification has had the benefit of experienced supervision and assistance as the student made the transition from principles and theories to the practical application of them in actual situations.

Finally, somewhere in the future, NIRA plans to develop some kind of certification or registration for industrial recreation technicians, by whatever name they may be known. These technicians or paraprofessionals are graduates of two-year programs, or those who have completed two years of acceptable course work, who thus have developed some specialty or skill and who have had scholastic internship in the capacity of a technical assistant. These technicians are important to the future of industrial recreation, particularly as programs become broader and more sophisticated. Many such technicians are already employed in the larger recreation programs; and some form of recognition of their knowledge and skills is desirable.

Licensure

Even stringent standards of certification are not the ultimate in professionalization. The ultimate is licensure. Certification encourages but does not require that all practitioners meet the requirements of professionalism. Licensure requires professionalism as a criterion of employment.

Licensure of industrial recreation administrators is an almost certain development, and it probably is not many years down the road. Licensure usual does, and should, follow only after stringent standards of certification have been developed. Licensure is a legal procedure which cannot be brought about by a profession but must be mandated by action of state or provincial legislatures. Licensure will come about only when the legislature of some state or province fully recognizes the social importance of industrial recreation and therefore appreciates the need to limit its administrators to those who have demonstrated their qualification by meeting all the requirements for certification. In view of the Nova Scotia conference of 1977, that province seems now to be one of the most likely originators of licensure for industrial recreation administrators. Among other likely pioneers of licensure are the states of California, New York, Ohio, and Texas. We feel comfortable in predicting that licensure will come though we know not where or when.

We can say something definite about worldwide aspects of industrial recreation, to which we address ourselves now in the final section of this book.

The World Scene

Industrial recreation is far from a monopoly of North America or, even, of the Western Hemisphere. Yet it is different in other parts of the world, as social and industrial conditions are different.

In Japan, though labor conditions are greatly different from those here, industrial recreation is particularly strong. In Japan, the individual's employment is usually a lifelong job with a single business or industry. It is critical there that the young person chooses wisely in selecting the industry in which the total working years will probably be spent.

Industries in Japan are very concerned about the welfare and happiness of their employees because of the long relationship and the recognized connection between employee morale and productivity. Japanese industries go to great lengths and expense in the effort to develop and maintain a high level of company loyalty. We will take time for two examples.

The Sony Company, in 1970, had 14,193 employees for whom it provided many recreational facilities and programs. These included: gymnasiums, tennis courts, a swimming pool, a recreational building for women, a club-style building for social functions, a playground for outdoor sports, a hotel and inn-style facilities by the sea, and a mountain resort available for use by employees and their families.[11]

At the same time, the Toyota Motor Company has a somewhat similar program for its 38,500 employees. Toyota provides free bus transportation to and from work for employees, and it pays for the gasoline used by workers who commute in their own cars. The company maintains a 344-bed hospital with twenty doctors in attendance. Toyota also provides a plentitude of sports and recreational facilities, including: swimming pools, meeting halls, tennis courts, golf courses, baseball diamonds, and rugby fields. Free classes are offered in the various activities. For vacationing employees, Toyota operates a string of mountain and seaside resorts which charged employees, in 1970, the equivalent of approximately $1.40 per day, including meals.[12]

There are some obvious differences between industrial recreation in Japan from that of this hemisphere. In Japan, practically all the expense is born by the employer, who is largely motivated by the desire to increase company loyalty and productivity. Under those circumstances, naturally, the emphasis is upon paternalism rather than upon democratic participation in decision making.

From what the authors have learned—but they make no claim to exhaustive research in this vast field—most foreign nations are prone to emphasize the health aspect of industrial recreation. For example, at the SABA Works in Villigen, in the Federal Republic of Germany, sports is a compulsory subject for apprentices. The wives of the men working at SABA are also allowed and encouraged to take part in gymnastic hours. The sports sessions are conducted by gymnasts and sports teachers which the firm employs.[13]

Of countries not in the Western Hemisphere, England appears to be most nearly similar to America in the development of industrial recreation. We believe that the employee club of GKN Screws & Fasteners is quite representative of the larger English employee associations. Of course, the existence of employee associations is significant in itself.

The GKN employee club dates from 1894. In 1977, it had about 3,000 full-time members. The GKN employees club director is a professional and a member of the Recreation Managers Association. The general policies of the GKN club are developed by the executive committee, which consists of three members appointed by the company—the Administrative Director, the Chairman, and the Secretary-Treasurer—and six full members of the club.

The club has a fine sports hall, built in 1974 and owned by the company. Its facilities and grounds provide for a considerable variety of team athletic sports and some social activities. It has a bar and a concert room seating 300 persons and used for dinner-dances, dances, concerts, social evenings, and dancing classes.

The GKN dues, 1 English pound per annum in 1977, provides a considerable sum of money, and there is some but apparently not a lot of other club-generated revenue. Obviously the club must depend to a very considerable extent on company largesse. The club does organize foreign vacation travel and weekend trips to holiday resorts.[14]

Although comparisons are risky with such a limited sample as we have and in the face of wide variations, we believe that the following generalizations are reasonably accurate.

1. Industrial recreation in England tends to be more democratic than in most other countries but is somewhat less so than in America.
2. Foreign industrial recreation is more dependent upon company support and is much less self-supporting.
3. The overwhelming emphasis in foreign countries is upon physical sports, team competition, and physical fitness.
4. Aside from some travel and music, there is very little attention paid to cultural aspects in most foreign programs.
5. Foreign programs appear to pay little or no attention to educational services for members.
6. Apparently few if any foreign programs concern themselves with their communities or community service.
7. Among foreign programs, we have found no reference to employee services of any kind. Certainly foreign clubs have not yet undertaken the large number of services common to American programs.
8. There appears to be markedly less emphasis upon family participation in some foreign programs. Some do include family members and family activities.
9. Industrial recreation abroad is considered largely as a fringe benefit and is provided in the expectation of deriving company benefits exceeding the costs.

Thus, while we see industrial recreation as a worldwide development, there are significant differences between foreign programs and those of this hemisphere.

If we are correct in our belief that the worldwide trend of history is toward greater democracy and the importance of the individual citizen—and we strongly believe we are correct—then, in time, industrial recreation abroad will become more democratic, more concerned about the development of the whole person, more family centered, and more community oriented. Yet only time will tell whether the programs we know will be seen as the pioneer examples which will influence the whole world.

Interest in Leisure

We do know that there is worldwide interest in the growth and use of leisure time and in leisure activities. There was recently an "International Congress of Leisure Activities in the Industrial Society" held in Brussels, Belgium. It was attended by Frank Flick and Arthur L. Conrad, J.D., CIRA, both of the Flick-Reedy Corporation. Conrad recounts that Flick and he found themselves "in a most peculiar position in Brussels when we stood almost alone with the Soviet Union on the need and necessity for work." They and the Soviet representatives differed markedly in some other respects, but they were in real agreement "about the balanced individual in society—one who works and plays."[15]

The worldwide interest in leisure finds expression in the well-endowed Van Cle Association Belgium. It has as its objectives the "taking and realizing in Belgium all kinds of initiatives at the service of humanbeing [sic] and the community in the field of free time, leisure activities, quality of life, promotion of culture and education, promotion of scientific research in the above areas."[16]

The Van Cle Association Belgium is working to develop a worldwide network of leaders interested in the problems and promises of growing amounts of leisure in the industrial world. Its letter of 8 August 1977 announced the Van Cle World Association "in course of foundation."

Regardless of the success of the Van Cle Association in the founding of a world organization, that association is symptomatic of the direction in which the world is heading. Assuredly it is one more straw in the wind indicating the growing importance and ultimate significance of industrial recreation.

Notes

1. *The Untapped Potential: Industrial Recreation.* National Industrial Recreation Association, Chicago, n.d.
2. "Determination of Attitudes of Employees and Management of Canadian Corporations toward Company Sponsored Physical Activity Facilities and Programs." University of Alberta, 1974.
3. Company Profile, *R.M.* September 1975, pp. 28–29.
4. *Business Week.* 13 February 1978.
5. The *Wall Street Journal.* 6 July 1978.
6. *The Untapped Potential.*
7. "Monthly Vital Statistics Reports: Births, Marriages, Divorces and Deaths for 1977." National Center for Health Statistics, DWHE Publication No. (PHS) 78–1120, 13 March 1978.
8. Lee Frankel and Alexander Fleisher. *The Human Factor in Industry.* (New York: The Macmillan Company, 1920), p. 288.
9. Lawrence Lemme. "The League of Federal Recreation Associations." *R.M.* February 1977, pp. 37–39.
10. "Government/Industry Team Promotes Employee Recreation." *R.M.* February 1977, pp. 17–19.
11. Sony administrator to Robert Wanzel.
12. "Japanese Labor's Silken Tranquility." *Time.* 5 October 1970.
13. "SABA." *Sportbild.* 1972. Edited by H. Ockardt, Bade Godesberg, German: Iner Nationes.
14. "A Look at Employee Recreation in England." *R.M.* April 1977, pp. 24–25. The article is a letter to Miles M. Carter, CIRA, from a colleague at GKN Screws & Fasteners, Ltd.
15. Arthur L. Conrad, J.D., CIRA, to Theodore Wilson, 29 December 1977.
16. Van Cle Association, Circular letter WA. 3, Grote Markt 9, 2000 Antwerpen, Belgium. The literature and other information about the Van Cle Association were provided by Arthur L. Conrad, J.D., CIRA.

Index